GOLD RUSHES

by Tony Hollihan

© 2001 by Folklore Publishing
First printed in 2001   10 9 8 7 6 5 4 3 2 1
Printed in Canada

All rights reserved. No part of this work covered by the copyrights hereon may be reproduced or used in any form or by any means—graphic, electronic or mechanical—without the prior written permission of the publisher, except for reviewers, who may quote brief passages. Any request for photocopying, recording, taping or storage on information retrieval systems of any part of this work shall be directed in writing to the publisher.

The Publisher: Folklore Publishing
Website: www.folklorepublishing.com

National Library of Canada Cataloguing in Publication Data

Hollihan, K. Tony (Kelvin Tony), 1964–
   Gold rushes

(The Legends series)
Includes bibliographical references.
ISBN 1-894864-01-8

   1. Gold mines and mining—Canada—History—19th century. 2. Gold mines and mining—United States—History—19th century. 3. British Columbia—Gold discoveries—History. 4. California—Gold discoveries—History. 5. Klondike River Valley (Yukon)—Gold discoveries—History. 6. Black Hills (S.D. and Wyo.)—Gold discoveries—History. I. Title. II. Series: Legends series (Edmonton, Alta.)
FC3820.G6H64 2001   971.1'02   C2001-910136-0
F1088.H64 2001

Photography credits: Every effort has been made to accurately credit the sources of photographs. Any errors or omissions should be directed to the publisher for changes in future editions. *Photographs courtesy of* Bancroft Library, University of California, Berkeley (p.23; p.25; p.28; p.46;); Centennial Archives, Deadwood Public Library (p.193); Glenbow Archives, Calgary, Canada (p.88, NA-674-10; p.116, NA-674-33; p.124, NA-674-51; p.129, NA-674-31; p.140, NA-3050-1; p.143, NA-674-48; p.213, NA-1466-5; p.221, NA-2615-10; p.235, NA-1052-1; p.267, NA-2067-1); Image Club Graphics (p.7) Library of Congress, Prints & Photographs Division (p.36, LC-USZ62-104557; p.64, LC-USZ62-12; p.169, LC-USZ62-78505; p.190, LC-USZ62-89966; p.198, LC-USZ62-95040);National Archives of Canada (p.9, C-19424; p.106, PA-125990; p.114, C-88930; p.132, C-88917; p.214, C-28599; p.231, C-28645; p.241, PA-13413; p.253, PA-17115; p.255, C-14477; p.259, PA-168973; p.263, PA-168981;); National Archives, Still Pictures Branch, College Park, MD (p.165, 999-ANSCO-CA-10; p.177, 77-HQ-264-801); National Library of Canada (p.160, C-14045); *The American West in the 19th Century*, by John Grafton,1992, Dover Publications (p.31; p.40; p.54; p.59);Vancouver Public Library, Special Collections (p.95,VPL 22860; p.237,VPL 9758);Wyoming State Archives (p.203, 24611); Yukon Archives (p.223, Skookum Jim Oral History Project Collection, 88/58R #33; p.233, Canadian Museum of Civilization Collection, #826; p. 247, McBride Museum Collection, #3782; p.250,T.R. Lane Collection, #1391; p.265,Vancouver Public Library Collection, #1991).

PC: P5

Contents

Introduction 4

Chapter 1
California (1849) 14

Chapter 2
The Fraser River (1858) 66

Chapter 3
The Cariboo (1862) 110

Chapter 4
The Black Hills (1876) 164

Chapter 5
The Klondike (1897) 210

Notes on Sources 270

∞

For Jackson, who shares the spirit of adventure.

INTRODUCTION

"THE NEXT EL DORADO!" the newspaper headline barked. It could have been the daily rag from New York, Melbourne, London, Halifax...heck, *any* city. And the date of the newspaper could have fallen in almost any decade of the second half of the 19th century. The plain truth of it was that if a fellow had a nickel for every time that phrase was printed during those years he wouldn't have had to prospect for his fortune. Yet, the magical draw of gold was that he probably would have anyway. He'd sell all he owned, and travel great distances through vast, unknown territory, for a crack at filling a pan with gravel scooped from a creek bed. He'd sit or tunnel for hours bent like a stunted jack pine until even stretching brought no relief. He'd cover the exposed parts of his body with charcoal for protection from the sun and insects. He'd live off food that months before he wouldn't have fed to his dog. And the odd time he might complain, cursing the creeks and the shafts and the bugs and the food and the...for he was only human. But leave? Not until the gold fever in his blood had run its course.

What was it about gold? It tempted millions, seduced hundreds of thousands, fulfilled the dreams of thousands, and left tens of thousands more penniless. The odds of being in a position to bathe in champagne, to buy all the fresh eggs in town, or to leave a can of gold dust at the door for use by unfortunates, in other words, the odds of striking it rich, were not in the prospector's favor. Folks knew this, but it didn't matter. There was no vaccination for gold fever.

It was the mere possibility of fortune that lured prospectors to the gold fields. It was a 19th-century lottery, a true egalitarian exercise, one in which everyone who participated had the same shot at hitting pay dirt. It didn't matter if a fellow arrived with a pick and shovel strapped to his back or if he had servants to carry the load. Muscles, calluses, and sweat were what brought reward…and a little luck. Sure, there were other reasons to work up the stake necessary to prospect. There was the sense of mystery that surrounded the enterprise and enticed the adventurous. Occasionally there were economic depressions that pushed unemployed men from their homes in stagnant cities to the gold fields where at least they could work claims. But the one constant, the one driving force was the infectious belief that whispered, "Well, why not me? Why shouldn't I be the one to make a fortune?" It transcended rational thought, but then again there was little that was rational about a gold rush. From the reasons for prospecting, to the towns that sprung up in the diggings, to the lifestyles of those who lived in those towns or along the creeks, there was precious little about a gold rush that an observer might mistake as having its origins in sound contemplation.

This book explores the five great North American gold rushes of the second half of the 19th century: California (1849), the Fraser River (1858), the Cariboo (1862), the Black Hills (1876), and the Klondike (1897). While there were other rushes in North America (and more than a few on other continents) during the period covered, none fired fevers as did these five. "Pikes Peak or bust" became a common cry of those destined for the gold diggings in the

Cherry Creek area of present-day Colorado. By the late spring of 1859 there were as many as 10,000 in Mountain City [Central City], but within a few months only a handful remained. It wasn't a prospector's rush because little gold was panned there. Most of it came from deep shafts, burrowed by companies with deeper pockets. Virginia City (Nevada), Leadville (Colorado) and Tombstone (Arizona Territory) came to boisterous life with the silver rushes. Despite population booms, and fortunes and streets every bit as dangerous and exciting as those in the gold towns of San Francisco, Barkerville and Deadwood, the silver rushes never really caught the public's imagination. It was gold and gold alone that made men's, and occasionally women's, hearts race.

The gold rushes shared many characteristics. The most significant was the type of mining. The lifeblood of the rushes was placer mining. Placers were gold-bearing deposits of gravel in creek beds and each of the rushes explored in this book had their origins in placer discoveries. Because of the placers, claims were first staked along creeks and rivers. Usually it was only a matter of weeks or months at the most, before all the valuable real estate was occupied. Those who arrived after that time were forced to prospect farther afield, work another's claim or buy a claim. It's not difficult to see that most of the money made by prospectors in a rush was made by those first on the scene.

Prospectors came with pick, shovel and pan, with the pan being the most important of the mining tools. The pan (*batea*) was a Mexican invention, originally a wooden bowl with gently sloping sides. Later it was made of metal, usually about 16 inches across. When used, the pan was filled about three-quarters full with gravel from a creek bed and slowly shaken or swirled just below the water's surface, so that the heavy material, including gold, could settle at the bottom of the pan. Rocks and debris were removed by hand, while the other sediment was swept away by the stream. Eventually any gold in the gravel became visible and was easily removed by hand. An average-sized pan took about 10 minutes to wash out.

Other tools popular along the creeks were the sluice-box and the rocker. The sluice-box was two-sided trough, usually four to six feet in length, but occasionally as much as double that, and up to 12 inches wide. It had a smooth opening and a riffled (ridged) section along the bottom from its middle to the far end. The sluice-box was placed in the stream with the opening slightly higher than the exit. The miner fed the upriver opening with gravel. Heavier sediment was trapped against the downstream side of the riffles. When the riffles were full of sediment, they had to be cleaned by panning. A one-man operation usually panned a yard or two of gravel a day.

The rocker, sometimes called a cradle because it resembled one, was similar to a sluice-box. It consisted of a trough with riffles on a downward sloping base and some holes at the far end. A box covered with a screen or mesh rested on top of the trough. A handle protruded from one corner of the box. Under the base was a set of rockers on which the unit rested. Gravel was thrown onto the screen and water was poured over it. As the cradle was rocked, the gravel was broken up, slipped through the mesh, was swept along the base with the water, and eventually out through the holes at the end. Again, the heavier sediment was trapped against the riffles and, once they were full, the sediment was cleaned. While a lone prospector could work a rocker, it was most efficient if three operated it. One man dug the gravel, one brought it to the rocker, and one poured water and rocked it. Allowing at least five yards of gravel to be washed per day, the rocker was a more efficient machine than the sluice. Because it was also portable, the rocker was also the more popular of the two.

Shafts and tunnels were also commonly used in mining but they were hard, backbreaking operations to build. Holes were sunk into bedrock, where gold-bearing gravel settled. Once a miner struck bedrock, he stood at the base of the shaft and filled buckets with gravel that his partner then hoisted out with a windlass. The gravel was washed out with a rocker, and subsequently cleaned with a pan. If the gravel proved to be gold-bearing, the miner would follow the "lead" by digging horizontally. In California, where the technique was first employed, it was referred to as coyote digging because the shafts looked like coyote dens.

Once the placer deposits and the tunnels were played out, which usually took about two or three years, the nature of mining in the gold rush areas changed. By that time, most miners were more than willing to unload diggings

Introduction

Windlass at the Barker claim, Cariboo Gold Rush

they could no longer work, so claims began to be concentrated in the hands of large companies, who were just as willing to buy them. These companies weren't interested in the low-key and inefficient operations of independent miners. They looked to the quartz that veined the hillsides. If it had a milky white or clear appearance, it was believed to contain the Mother Lode, which was the source of gold in the placer deposits. The problem was that mining the quartz (hard rock mining, as it was called) was expensive. It required heavy equipment and laborers, and only big companies could afford either. When they moved in, the prospectors moved on and the rushes were more or less over.

The gold rushes shared other characteristics. They each demanded great courage from their participants. Whether it was the thousands of miles that had to be conquered to reach gold country or the often rough conditions prospectors discovered upon arrival, the one thing that can be said about a gold rush was that it was not for the faint of heart. Each gold rush also had boom towns that exploded from small populations to tens of thousands of people. The lifeblood of such towns was in supplying the miners' needs. Everything from food and gear to booze and women were available. There was little to differentiate the saloons, hotels and brothels in Yale, Dawson City or Deadwood. And in each town, it was those who provided the services who had the best chance of actually striking it rich. During a rush, store owners, restaurateurs and saloon keepers always had a new or ready supply of patrons all too willing to part with their poke (stake). Those who did well for themselves also included professional gamblers, who were as common in gold rush country as the piles of manure that messed the streets. They found that a man with gold was an easy mark.

As similar as the gold rushes were, they also had their distinct features. The most marked difference was a political one. The Fraser River, the Cariboo (both in British Columbia) and the Klondike (Yukon) rushes occurred in Canada, while those in California and the Black Hills (South Dakota/Wyoming) were in the United States. In many respects, miners treated borders like the invisible lines they were. It's likely that most prospectors weren't even aware that they were entering Canada when they stampeded north. They learned soon enough. The Canadian gold rushes were mostly orderly affairs, at least in comparison to the crime, violence and vigilante justice that were common features in the United States. In British Columbia the law was laid down by Governor James Douglas and enforced by Judge Matthew Baillie Begbie, while in the Yukon the North-West Mounted Police were on patrol. Their efforts ensured that miners could actually work their claims and enjoy the fruits of their efforts without having to worry about either their pokes or their lives. It's not merely a coincidence that American criminals and outlaws such as Persimmons Bill or Joaquin Murieta were not to be found north of the 49th parallel.

There were other differences. While women participated in all the gold rushes, they were mostly employed in dance halls, as domestic help or as prostitutes. In California, however, Mexican prospectors often arrived with their wives and families. There was a similar pattern in the Klondike, where the sight of a family rarely raised eyebrows. But it was only in the Klondike where women participated as miners with any regularity.

The sparks that set off each gold rush were also largely distinct. Only two of the rushes, the Cariboo and the Klondike, were actually launched by prospectors searching for gold. In California, gold was accidentally discovered as a sawmill was being constructed. Along the Fraser River, the Hudson's Bay Company traded with the Natives for gold, and the rush was set off when one of their ships arrived in San Francisco laden with the ore. In the Black Hills a military reconnaissance expedition discovered gold.

In whatever manner the gold was stumbled upon, it was always the case that Natives peopled gold country before the prospectors arrived. In none of the rushes did the Natives fare well; in fact, they all lost most of the land they occupied. However, the process of stripping them of their territory was rarely similar. Sometimes it depended on the response of the Natives. The Sioux in the Black Hills took to arms and fought the white interlopers before they were finally forced to relinquish title. In the Klondike, the Hän were simply relocated, having been told that the little piece of land they were given as compensation was all they'd ever get in return for what was taken, at least until the region was cleaned of its gold. Sometimes the Natives participated in the rush. In California and along the Fraser River they were known to mine for gold. In the Yukon they supplied meat and furs. But in all the rushes, many Natives were victims of prejudice and alcohol, surviving only by begging or prostitution.

Miners rarely took any gold home with them after their gold rush adventure. Some were bitter about it. The mining had been physically demanding work done in stressful conditions. Loneliness, hunger, restless nights, and fear, all experienced in strange and often foreboding locales, took their toll. The sentiment was well captured in J. A. Stone's song, *The Lousy Miner*.

> *I've lived on swine till I grunt and squeal,*
> *No one can tell how my bowels feel,*
> *With flapjacks swimming round in bacon grease.*
> *I'm a lousy miner,*
> *I'm a lousy miner; when will my troubles cease?*
>
> *Oh, land of gold, you did me deceive,*
> *And I intend in thee my bones to leave;*
> *So farewell, home, now my friends grow cold,*
> *I'm a lousy miner,*
> *I'm a lousy miner in search of shining gold.*

For others, gold rush country was a place of dreams, even if the shining gold was never discovered. In a gold rush, people had the opportunity to test their mettle, to determine what it was that they were made of. There were also those who forged lifelong friendships and who rued the day they had to leave. It was an attitude well summed up in James Anderson's *Farewell!*

> *Cold Cariboo, farewell!*
> *I write it with a sad and heavy heart;*

You've treated me so roughly that I feel,
Tis hard to part.

Farewell! a fond farewell
To all thy friendships, kindly Cariboo!
No other land hath hearts more warm than thine,
Nor friends more true.

Apparently, it wasn't always about the gold.

CALIFORNIA
(1849)

IT WAS THE SPRING OF 1849 and a clipper ship was readying to slip out of Salem harbor. It just as easily could have been Boston, Baltimore, New York or any port on the United States' eastern shoreline between Maine and Louisiana. They were all abuzz. There wasn't a berth to be had, though many were willing to pay more than the $325 usually charged for a ticket. Some latecomers found a ship's captain who was willing to squeeze in another passenger or two. And despite facing five or six months in a cramped space just large enough for a hammock strung from walls never meant to quarter passengers, they considered themselves fortunate. But a man headed for California's gold fields was willing to sacrifice a lot to reach the diggings without delay. Such men were soon to discover that there was plenty more sacrifices to be made during the long journey around South America's Horn.

As the dock workers threw off the mooring lines, the ship's deck brimmed with passengers. To a man they serenaded those well-wishers who had come to see them off.

The song of choice was an adaptation of Stephen Foster's *O Susannah*, first sung in Salem months earlier.

> *I came from Salem City,*
> *With my washbowl on my knee,*
> *I'm going to California,*
> *The gold dust for to see,*
> *It rained all night the day I left,*
> *The weather it was dry,*
> *The sun so hot I froze to death*
> *Oh brothers don't you cry!*
>
> *O California,*
> *That's the land for me!*
> *I'm bound for San Francisco*
> *With my washbowl on my knee.*
>
> *I soon shall be in Frisco,*
> *And there I'll look around,*
> *And when I see the Gold lumps there*
> *I'll pick them off the ground.*
> *I'll scrape the mountain clean, my boys,*
> *I'll drain the rivers dry,*
> *A pocketful of rocks bring home—*
> *So brothers don't you cry!*

Spirits were high! They remained so for some time, at least once everyone got their sea legs. Sailing in the North Atlantic in spring was rarely enjoyable for the first few weeks, but green faces eventually gave way to healthy pink tones and soon the railings were free from the sweaty grasps of those leaning over them to discard undigested meals.

Soon all settled into a routine that served well enough for the first couple of months of the voyage. Games were popular, especially if they were suited for gambling, and folks often played checkers, backgammon or poker. Some took to reading, while others listened intently to stories. Those with the foresight to pack booze drank it. Plenty had, and perhaps that contributed to the practical jokes that were regular fare. A most popular one was to place a bucket of seawater atop a door. Watching a fellow sputter and shake off the wet surprise provided plenty of laughs, but guffaws turned to lowered eyes and hasty exits when the captain was unfortunate enough to be the victim. The jokesters weren't at a loss for pranks, however. One passenger fancied himself a fisherman, spending many long but fruitless hours near the stern of the ship with his line trawling along. When there was a commotion near the bow of the ship—contrived, but unknown to him—he tied his line and scurried off to investigate. Upon returning, he discovered that his line was tight. Excited, he called to all. "We'll be eating tuna tonight!" Amidst the cheers of his fellow passengers, he reeled in the line, eventually depositing a chamber pot on the deck.

After a few months at sea, enthusiasm began to wane. Stories told two or three times were far less interesting. Folks took to packing and repacking their gear. Others spent time guessing the ages of fellow passengers. Soon even that was no longer entertaining, so they spent long hours watching the rolling waves. Some turned to diaries, drawing and music, occasionally discovering talents they didn't know they possessed. All enjoyed a brief respite when the vessel crossed the Equator and the passengers were granted an audience with Neptune, whose appearance (or some variation thereof) was a time honored tradition among sailors. It was followed by a short layover in Rio de Janeiro, where stores were restocked. The new supplies didn't come soon enough and rarely lasted long enough.

The passengers might have put up with boredom, the close quarters, or any inconvenience, if they'd had proper meals to eat. In the weeks after leaving port, food presented little problem. Lobscouse, a stew made from potatoes, salt meat and hard tack, was tackled with zest. A fellow could even clean a bowl full of hushamagrundy, because its ingredients of turnips, parsnip and ground cod weren't too foreign. But as supplies dwindled and ship's cooks were forced to become more creative, meals had to be eaten with less attention given to appearances and ingredients. Dried goods, such as beans, stored well, and they became standard fare as the voyage progressed. Unfortunately, the water needed to cook them had a shorter shelf life. Often barrels of water were cracked open and found to be the breeding grounds for countless insect species. But there was no throwing it overboard; it was a precious commodity that had to be used. Vinegar was liberally added to kill the bugs, followed by a few dollops of molasses to kill the taste. After drinking switchel, as it was called, pink faces were soon green again.

Like most of those bound for California by sea, those on the vessel out of Salem were well into the switchel by the time they reached Cape Horn, which was an unfortunate coincidence. The stretch around South America's southern tip was white with stormy waves the likes of which could turn even an experienced captain's stomach. The passage was so rough that many preferred to be sick in their quarters rather than risk a staggered journey above decks to the rail. Added to that was the bone-chilling cold of a winter in the Southern Hemisphere. But the worst of it all was the knowledge that San Francisco was still months way, that is, if they were lucky enough not to find themselves becalmed, or worse still, forced to retreat before a sudden Pacific storm. The days of singing *Oh California* seemed distant indeed.

Those swept along to California in the rush of '49 were hundreds of years removed from the first men who had traveled to the Pacific shores of North America in search of gold. Columbus was still a squatter when stories of the precious metal began to grip the imaginations of Europeans. In the early 16th century, one of the first of those stories christened the *terra incognito* California. Garcia Ordoñez de Montalvo wrote about an island of that name inhabited by a warlike tribe of women, the Amazons. They carried weapons and wore armor of solid gold, which was hardly surprising given that Montalvo claimed the only metal to be found on the island was gold. Those anxious to claim the riches of California for themselves did so at their own peril because the Amazons kept griffins as pets, and the creatures fed on men who fell within their carnivorous reach.

Such dire warnings did little to deter fortune seekers, especially when other stories emerged apparently confirming the speculated riches of the region. Men's eyes glazed over as they listened to the tale of El Dorado, the golden man. The story told of a Native tribe that performed an annual rite that saw its chief covered in a layer of gold dust. There was some truth to it. The Chibcha of present-day Colombia had practiced the ritual, though not for many years when Europeans arrived. As the story became a legend and was further embellished, the gold-coated chief became a golden man and eventually a golden city. Tales of El Dorado seemed to be corroborated by rumors of seven golden cities supposedly built on high plateaus, their brilliance matched only by that of the sun itself.

The first Europeans to chase such stories were the Spanish. They knew something of gold in the New World. In the years following their arrival they had filled their imperial coffers with plunder taken from the Aztecs and Incas. The wealth of those peoples only seemed to validate the stories that suggested riches were to be found in the lands to the west. Throughout the first half of the 16th century, expeditions were dispatched, the most prominent of which was that of Francisco Vásquez de Coronado. In the summer of 1540 Coronado set off in search of the seven golden cities. Anchoring somewhere on the southeastern coast of North America, his expedition pressed inland to the northeast as far as present-day Kansas. A section of the party split off and went west to what became California. There they found fellow explorer Francisco Ulloa decimated by dysentery. Ulloa had found gold, but apparently he buried it rather than allow others to take it. Others in the Coronado expedition set off to explore the Sea of Cortez. They hoped they might find China, believing that the seven golden cities might be in Gran Khan's territory. Instead, they sailed up what would be called the Colorado River, as far as the southern reaches of Arizona.

Coronado didn't find the seven golden cities, and save Ulloa's fortune, neither did he find gold. The news of Coronado's failure did not dampen the enthusiasm of others, especially the English, who were not inclined to believe Spanish assertions about the lack of gold in the region. The first English arrival was likely the adventurer Sir Francis Drake. He was as interested in pirating Spanish vessels and finding the Northwest Passage as he was in exploring for gold, but as he sailed by California in the late 1570s he took the time to drop anchor and claim the region for Elizabeth I. No one knows where he landed or whether he actually saw California gold. Even had he returned to England with the hold of the Golden Hind filled with the ore, it is doubtful that his arrival would have set in motion any mass exodus for the gold territory. It was too far away and travel around Cape Horn through the Straits of Magellan was far too dangerous.

Perhaps it was inevitable then, that California remained within the Spanish sphere of influence. Throughout the 17th century, the Spanish expanded their holdings northwards into the region, first establishing a colony at San Diego in 1602. Throughout the remainder of the century and into the following one, little changed other than knowledge of the region's geography. By the 1680s, for example, it was finally determined that California was not an island. By the mid-18th century renewed interest from the English and the discovery of a Russian presence, whose fur traders were pressing in from the north, spurred the Spanish into renewed interest in the region. Officials in Mexico directed priests to establish missions along the Pacific coast to the north. Often colonies were established adjacent to gold deposits, their whereabouts having been revealed by local Natives. Natives usually worked such mines and, though they did not share in the profits of their labors, they seemed to get along well enough with the Spanish.

In the last decade of the 18th century, the fur trade in the Pacific Northwest began to flourish when it was learned that there was an insatiable market in the Orient for the region's sea otter pelts. British traders arrived in increasing numbers and Spain reacted with a show of force. In 1789, they seized two British trading vessels in Nootka Sound (on the western side of what is today known as Vancouver Island) and took them to Mexico. The action led to a heated international affair, almost bringing Spain and Britain to war. As the two countries twirled their diplomatic dance, an unexpected presence slipped onto the scene. In 1792 Robert Gray, an American, sailed his ship, the *Columbia*, from Boston to Nootka, hoping to cash in on the fur frenzy. On his journey, he chanced upon the mouth of what was to become the most important river on North America's western edge, the Columbia (named after his ship). The discovery gave the United States a claim to the Pacific shoreline that they would use to expand their empire to the north and south.

By the 1820s territorial rights were beginning to crystallize. Britain and the United States reached what would ultimately prove to be an unsatisfactory compromise in 1818, one that opened the Oregon Territory west of the Rockies to citizens of both countries. The agreement was the subject of debate for another 30 years, but it recognized that the British had no claim south into California. Neither did the Russians, a fact recognized a few years later when Russia also agreed to abandon all claims south of 54° 40'. But California was not American territory. The Treaty of Florida Blanca, signed in 1819, set the boundary between Spain and the United States at the 42nd parallel, the northern border of present-day California.

Within a few years, Spain was no longer a factor in the Pacific intrigues. In 1821 Mexico won its independence from Spain, but it would not retain California for long. In the 1840s an expansionist craze swept America. When John L. Sullivan, a New York journalist, wrote of his country's need to fulfill its "manifest destiny to overspread the continent allotted by providence..." he struck a responsive chord. James Polk had recently been elected president on an expansionist platform and he moved quickly to fulfill his promises. In the southwest, hungry American eyes looked longingly at northern Mexico. Only possession would satisfy whetted appetites and, in early May 1846, the two adversaries collided north of the Rio Grande. War raged for the better part of two years until February 2, 1848, when the Treaty of Guadalupe Hidalgo was signed. For $15 million, the United States purchased present-day

Nevada, Utah, huge chunks of New Mexico, Arizona and Colorado, and the territory that would become the state of California.

The signatures on that piece of paper changed the lives of tens of thousands of people. However, one man's life was to be dramatically transformed a mere handful of days before the treaty was signed. His name was John Sutter. A Swiss born in Germany, Sutter arrived in California with empty pockets in 1839. What he lacked in cash he made up for in audacity. His warm out-going ways had served to further his business efforts on many an occasion, and they were just as effective with the Mexican territorial governor. It's likely such a meeting would never have taken place had Sutter knocked on the man's door and told the truth of his spotty entrepreneurial record and the European warrant for his arrest. Instead he passed himself off as a retired captain in the Swiss Guard of King Charles X of France. The meeting went well; he left with a grant of 150,000 acres near present-day Sacramento, which he called Nueva Halvetia (New Switzerland). He became a Mexican citizen and had no trouble settling into the easy life enjoyed by prominent Spanish-Mexican families. Servants worked his rancho, tended his herds and picked his grapes. The rancho was well secured, with towers, turrets and brass cannons so locals took to calling it Sutter's Fort.

Sutter had grandiose dreams and envisioned his rancho as the centerpiece of a thriving community. To bring his feudal dreams to fruition, he built a sawmill that was to provide the necessary building materials. He hired James Marshall, one of the many visitors to Sutter's Fort, to build and run the mill. Marshall headed up the American River and found a suitable location some 45 miles away, which he dubbed Coloma. By January 1848 the mill was almost completed and only one problem remained to be solved. The tailrace

Johann Augustus Sutter (1803–1880), a German-born Swiss, came to the West in the 1830s looking to make his fortune in farming and ranching. He acquired land from Mexico and built a sprawling ranch on 150,000 acres at the junction of the Sacramento and American rivers. His domain included a gristmill, tannery and distillery, in addition to the sawmill he hired James Marshall to construct at Coloma. Marshall proved to be a better builder than a surveyor. Unfortunately, the site he had chosen lay three miles outside Sutter's holdings. Sutter did lease the territory, but he was eventually ruined by the legal expenses of defending his claim.

below the wheel was too shallow to adequately power the saw's blade. Marshall tried increasing the flow of water through the tailrace in an effort to deepen the channel. On the morning of January 24, he set out to evaluate the effort. His attention was quickly drawn from that task when a glittering object in the water caught his eye. He scooped it up. It was a piece of gold about the size of a pea! Actually Marshall doubted it was gold; he had a gloomy outlook on life and believed that such luck was beyond his grasp. He tested it. Hammering proved its malleability and submerging it in lye had no altering effect. Doubts became disbelief. It was gold.

Marshall took the small nuggets (he had found more) back to Sutter's Fort where Sutter subjected them to more tests. He was soon convinced and set out with Marshall to the mill. Marshall wanted to ensure that Sutter actually found gold and to that end he had salted the tailrace by filling it with gold he had already discovered. Before the pair arrived at the mill, a son of one of the workers found the precious metal. The effect on Sutter was the same. When the boy rushed to greet them with his find, Sutter could only exclaim, "By Jove, it is rich!" There was, however, bad news awaiting Sutter. The mill had been built outside Nueva Halvetia. He set about trying to obtain title to the land. Hurried negotiations with the local Natives won him a three-year lease to the region.

Sutter had less success trying to keep the discovery a secret. His workers took to prospecting in their spare time, and it wasn't long before they found enough gold to allow them to abandon Sutter's employment. It was an understandable career change, given that a week's mining pocketed them the equivalent of a year's pay back east. By March, newspapers on the west coast were reporting the find, but the news piqued little interest. Marshall was not the first to pluck gold from a riverbed in California. In the

Millwright James Wilson Marshall (1810–1885) had no idea, when he built a sawmill for John Sutter, that his efforts to make the wheel more efficient would have any far-reaching consequences. When he spotted gold nuggets in the tailrace channel, he immediately took them to Sutter. Once Sutter had confirmed the ore was gold, he asked Marshall to keep his find secret. But the discovery of gold could not be kept under wraps, and within a few weeks the news had spread to Samuel Brannan, who quickly set up a store in the vicinity and made it his mission to attract prospectors from across the continent.

early 1840s geologists confirmed what folks had long dreamed and often known: there was gold in California. Folks didn't respond to the news, perhaps because it wasn't properly publicized. El Dorado wanted only a public relations man and it found it in Sam Brannan.

⁂

When dawn broke on May 12, 1848, no one had any reason to believe that California would ever be the same again. On that day Sam Brannan took up a position at the corner of a busy plaza in San Francisco. Though he was well groomed and finely attired, his odd show suggested that he might be just a little touched in the head.

"Gold! Gold from the American River!" he shouted at the top of his lungs.

And he had more than words to offer. As he called to the onlookers, he jabbed a bottle of gold into the air above his head. The glittering lure attracted a crowd, and brief seconds of quiet consideration soon gave way to enthusiastic chatter. Folks in San Francisco had been quick to dismiss the rumors of gold drifting west from Nueva Halevita. Many assumed the stories were nothing more than some new con designed by Sutter to attract workers or to fill his coffers. But here was the evidence. Within days most of the male population of the town had left for Sutter's Fort, and Brannan, surely, was wearing a great smile. He was about as crazy as a fox.

Samuel Brannan was a Mormon Elder, a prominent member of the religious community that was to play a significant role in turning the gold discoveries into a full blown rush. Mormon leader Brigham Young sent Brannan to California. He was assigned as the guide to a party whose task it was to collect tithes from Mormons in the region. Brannan was upset that he wasn't appointed leader of the expedition, but he had no problems with going to California. In fact, he had heated arguments with Young about the desirability of the church using California as its base of operations rather than Utah. Once he left for California, Young was glad to be rid of him. Rather than return to Utah, Brannan decided to remain in California. He was persuasive speaker and convinced other Mormons that a winter in Great Salt Lake was sure to compel Young and his followers to continue west. In the shadow of Sutter's Fort, he established the New Hope colony and, with his partner Charlie Smith, opened a store. With the news that Young would not abandon his chosen site, many of those living in New Hope left. Most who remained began working for Sutter. Among their number were the bulk of those who would build the Sutter-Marshall sawmill.

When Brannan learned of the discovery of gold at Coloma, he decided to ride out and investigate for himself. His own eyes confirmed the rumors. It didn't take long for Brannan to imagine the possibilities. Unlike Sutter, who saw a fortune to be made in keeping the gold a secret and mining it for his own benefit, Brannan knew that, as a merchant, his future prosperity rested on promoting the find. The arrival of prospectors would make his store a going concern. As he collected enough gold to fill a bottle, his mind raced with the possibilities: a warehouse at the Sacramento River Landing, a hotel in Sacramento....

However, a merchant couldn't make a fortune from the small numbers who came from San Francisco in the spring of 1848, despite their wild enthusiasm. Brannan needed immigrants, and to catch their imagination he began sending reports of the discovery of gold to eastern newspapers. It wasn't until August that any decided to print his stories.

If the names of Sutter and Marshall became famous by virtue of their gold discovery, then Samuel Brannan (1819–1889) became its public relations man. He devoted himself to attracting settlers and drumming up business for his many and varied business enterprises. A Mormon elder, he came to California via Utah charged with collecting tithes from local Mormons. When he confirmed the gold discovery, he immediately leased a house and set up a store at Mormon Island. He never intended to prospect for gold, but he did his best to advertise the discoveries so that others would come and take advantage of the goods and services he offered—for a price, of course. He made a fortune…and lost it all. Brannan was also head of the first Vigilance Committee, a group of 500 businessmen who patrolled the streets of San Francisco trying to stamp out crime.

Even then Brannan's claims were met with doubt. Where Brannan's word was insufficient for the outside world, Colonel Richard Mason's proved to be somewhat more persuasive. Mason was the United States military commander and acting Governor in California. In July he toured the gold country, visiting Mormon Island, Coloma, Weber's Creek and other as yet unnamed places. His report was thoroughly spirited. He noted that the region was alive with men and crowded with tents and shanties. Thousands of dollars worth of gold were mined with relative ease and little capital. One soldier reported that the money he had made in a week of mining was in excess of what he expected to make in five years of enlisted service. Perhaps this was an exaggeration, but Mason figured it was reasonable to suggest that one day's work could gain the equivalent of a soldier's monthly pay. But the coup de grâce of Mason's report was the glittering future he foresaw: "[a] small ravine... [has produced] $12,000 worth of gold. Hundreds of similar ravines, to all appearances, are as yet untouched."

Mason's report made its way back east by fall and was confirmed by similar accounts. On December 5, 1848, President Polk addressed the United States Congress and informed them of the discovery. "It was known that mines of the precious metals existed to a considerable extent in California at the time of its acquisition. Recent discoveries render it probable that these mines are more extensive and valuable than anticipated," assured the president. "The accounts of the abundance of gold in that territory are of such an extraordinary character as would scarcely command belief were they not corroborated by authentic reports of officers in the public service, who have visited the mineral district, and derived the fact which they detail from personal observation."

A presidential endorsement did much for California. Quickly eastern cities were alive with self-appointed experts who lauded the gold discoveries. In New York, one suggested, "The supply of gold was absolutely inexhaustible; and... one hundred thousand persons could not exhaust it in 10 or 12 years.... [Miners earned] from an ounce to a thousand dollars a day." Newspapers soon leaped on the bandwagon, spreading other eyewitness accounts. The *Washington Daily Union* speculated that "The gold in California must be greater than has hitherto been discovered in the old and new continents." When New York's *Literary American* quoted a San Francisco official as saying, "The streams are paved with gold—the mountains swim in their golden girdle—it sparkles in the sands of the valley—it glitters in the coronets of the cliffs," there was no stopping the rush.

The number of prospectors multiplied exponentially. The handful working at Sutter's sawmill in January became 4000 in July when Mason wrote his report and easily grew to 10,000 by year's end. After arriving at Coloma or Sacramento, prospectors quickly fanned out through the interior. Soon the rough boundaries of the gold fields—diggings as folks took to calling them—were sketched out. Sonoma was considered to be at the eastern boundary of the deposits and, during the height of the rush, few prospected past Lake Tahoe in the east. The northern frontier was marked by Lassen's Ranch on the Sacramento River, while it extended to Mariposa on the Merced River in the south.

Before January 1849 drew to a close, at least 10,000 had left from the east coast. A unique feature of the rush of '49 was that many of these argonauts (so called for the Greek legend of Jason and his ship the Argo that sailed in search of the Golden Fleece) traveled as members of joint stock

James Marshall was building Sutter's Mill when he made the glittering discovery that ignited the California Gold Rush. The mill was located at Coloma on the south fork of the American River, just beyond the borders of John Sutter's enormous ranch, Nueva Halvetia. Sutter, a German-born Swiss businessman and adventurer, settled in the region in 1839. He spent nearly a decade developing livestock and agricultural operations, oblivious that the surrounding land was laced with mineral riches. Yet the discovery of gold at the mill would be one of the bitter ironies of Sutter's life. When the news of the gold discovery became public, all the laborers in Nueva Halvetia were infected with gold fever. No one was willing to work for Sutter when gold mining promised to be much more lucrative, and his enterprise ground to a halt.

companies created just for the rush. In 1849, 102 such companies departed from Massachusetts alone. It was an effective way to gain a share in the diggings without investing too much capital since it allowed men to pool their resources. Investors split the profits from their ventures. Often these profits came not from discovered gold, but from the sale of items. Everything from bolts of cloth to prefabricated houses and toothpicks sold in San Francisco. Supplies such as these filled the cargo holds of the ships they chartered.

While the argonauts apparently knew something about making money, they knew as much about California as they did Timbuktu. As a result, some of the first entrepreneurs to profit from the glitter of gold were those with ready access to publishing companies. Late 1848 and 1849 saw a proliferation of gold-seeking manuals, travel guides and California fact books, such as *The Emigrant's Guide to the Gold Mines: Three Weeks in The Gold Mines, or Adventures with the Gold-Diggers of California, in August, 1848, together with Advice to Emigrants, with Full Instructions upon the Best Methods of Getting There, Living, Expenses, etc., etc., and a Complete Description of the Country, with a Map and Illustrations.* At 30 pages, it had the added bonuses of being a quick read and easy to pack for the journey! Many of the books outlined the gear a prospector would need. Upwards of $1000 was a good start to pay for it. Basic necessities included a woolen shirt, homespun flannel underwear, knee-length coat, jeans, trunks, felt-brimmed hat and water-proof bags to take it all in. Revolvers and bowie knives were also advised as necessary. More than one optimistic argonaut threw in an evening suit, which they fully expected to don once they struck it rich. Mining equipment could be purchased on arrival in California.

Properly equipped, the argonaut had only to find a way to get to gold country. Although his choices were not limitless, they were many. Usually money and time were the deciding factors. Most popular was the ocean voyage around the Horn. At anywhere from $250 to $400, a ticket didn't eat up too much of the argonaut's cash; however, it did consume plenty of his time. The journey was about 15,000 miles and even a half-decent ship could expect to take six months to cover the distance. In the heady days of the rush, however, many ships were of poor quality. Old ships best scuttled were brought out of retirement, so it was not uncommon to hear tragic stories of their sinking. Even if a fellow was fortunate enough to book a berth on a seaworthy vessel, there was no guarantee his trip would be enjoyable. Over 70 vessels of New England's whaling fleet were transformed into passenger ships; however, while planking could be removed to build quarters, the stench of whale blubber was there to stay. And even the best of ships had to pass through the Straits of Magellan, whose storms challenged the wiles of the saltiest of captains.

An argonaut could lop off a significant amount of time and discomfort from his trip if he decided to take a ship as far as Panama or Nicaragua, travel across the land barrier, and sail the rest of the way to San Francisco from Panama City. Time was of paramount concern to most prospectors, because all harbored the fear that they would arrive a little too late to cash in on the bonanza. Most went via Panama because the land route was shorter and better known (it was the route used by American mail steamers). A ship dropped them off at the mouth of the Charges River, where some tough bargaining with locals secured a guide to lead them upriver. After a 60-mile endurance contest that included challenging trails, exhausting heat

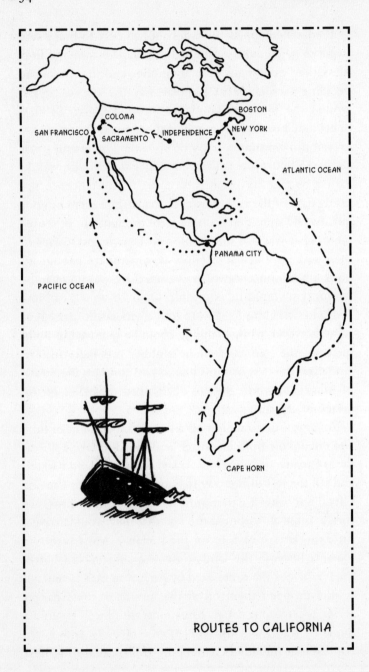

and, depending on the season, clouds of merciless, disease-ridden mosquitoes, the argonauts arrived at Panama City where they waited to board a ship bound for San Francisco. By 1855, a railway covering the route was constructed, but while it made the land trek a little more comfortable, it didn't save much time. In all, the trip took about three months and, at only $100 for the cross-isthmus journey, it left pocketbooks in a healthier condition.

Those uncomfortable with a lengthy sea voyage, and they numbered in the tens of thousands, surveyed the situation and decided that the most sensible route to California was by land. There were plenty of possibilities once a fellow made his way to the American midwest. He could head south and take the Santa Fe Trail to Santa Fe, continue on to San Diego on the Gila Trail, and link up with the Anza Trail to Monterey, California. If a northern route was preferred, the forty-niner could take either the Oregon or Mormon Trails, both of which eventually linked up with the California Trail and led right to Sutter's Fort. These routes weren't more than a few thousand miles, and an industrious argonaut could usually cover one of them in less than 100 days. The only expenses were a wagon and supplies; these were reasonably priced until demand exceeded supply, which didn't take long. And, because many of the cross-continental voyagers also traveled in companies, costs were shared and thereby further reduced. The only concern was that of safety. Folks had to be prepared for thirsty desert stretches, possible encounters with hostile Natives, and the dangers of veering from the trodden path. Little beyond keeping a loaded rifle close by could be done about the Natives, though they weren't a real threat for another decade or so. Most had heard of the grisly deaths of the Donner party, a westbound immigrant party that, in 1846, got lost in the

The discovery of gold started a stampede for California. Driven by promises of incredible wealth and fears that the best claims would be gone by the time they got there, forty-niners fell on top of each other trying to get to California as quickly as they could. It was a frantic exodus. There were barely enough ships on the Atlantic seaboard to transport hopeful prospectors to their Promised Land. The Oregon–California Trail was choked with thousands of covered wagons, practically forming a continuous line from Independence, KS to the gold fields. Currier and Ives captured the desperation of these 19th-century argonauts in the 1849 lithograph above. While no one actually took a rocket and few availed themselves of Rufus Porter's Aerial Locomotive to get across the continent, the need for fast transportation was so pressing that more than one man considered the possibility.

Wasatch Mountains and resorted to cannibalism in a desperate attempt to survive, so they were keen on sticking to the known trail. Nevertheless plenty died, mostly of cholera. Others ran out of supplies or saw their animals starve. Sometimes dead animals were just bad luck, but there were many cases of forty-niners burning the grasslands behind them so that the next wave bound for California would be delayed in arriving.

Undoubtedly, the best deal going to get to California was the one offered by Rufus Porter, founder of the *Scientific American*. He dreamed up an 800-foot long Aerial Locomotive, a hot air balloon powered by steam engines that would travel at 100 miles per hour and whisk forty-niners across the continent in three days. All that plus wine for only $50. Porter built his Aerial Locomotive, but couldn't sell enough tickets to make a go of it. Obviously there was a limit to the forty-niners' courage, if not their need for speed.

As for the three men, John Sutter, James Marshall and Sam Brannan, who were responsible for striking the spark that set this hurried rush ablaze, they each ultimately suffered a similar, tragic fate. Sutter spent a fortune trying to defend his land claims in court. Facing debt and irate creditors, he moved to Pennsylvania with his family where he died a poor man. Marshall enjoyed a brief reputation as a man with the golden touch, so he was often hired to find gold, but he never succeeded. He came to depend on a government pension, which was subsequently revoked when it was discovered that he spent most of it on booze. He also died in poverty. Brannan truly struck it rich. He bought out his partner in their Sacramento store for $50,000, and subsequently acquired large holdings in that city and in San Francisco. He owned a hotel, a newspaper and a bank. He abandoned his Mormon ways, unfortunately before his wife discovered that he had another wife back east (polygamy was permitted in the Mormon church at this time). His California wife filed for divorce and the settlement ruined him. He turned with increasing vigor to the bottle and died a pauper. None of the three lived to see the 1890s.

~

Many towns and cities were forged in the heat of the California Gold Rush. Places like You Bet and Rough and Ready sparkled only as long as there was gold to be found. Others, such as Sacramento and Stockton, enjoyed enduring longevity after the forty-niners were but memories. And then there was San Francisco. It was a city like no other, a boom town that exploded during the rush of '49 and continued to grow thereafter. The burg caught the popular imagination and, for many forty-niners, San Francisco and the gold rush were soon synonymous. This sentiment was well expressed in one ditty of the time:

> *The San Francisco Company,*
> *For San Francisco bound,*
> *Our barque is San Francisco too,*
> *The same name all around.*

Prospectors dreamed of gold while in New York, Melbourne, Shanghai or any of a thousand other places, but for most of them it wasn't until they stepped ashore at San Francisco that those dreams were within reach.

San Francisco had something of a history before the gold rush but little hinted at the great future that lay in store. Spanish authorities, keen on protecting imperial interests, had directed priests to establish a mission in the region in 1776. Seventy years later San Francisco was no longer a mission; it was controlled by Mexico and the population had stagnated at 150. In 1845 an American trader set up shop in Yerba Buena (as San Francisco was then named). He enjoyed enough success that, when the Mexican-American war erupted, it was evident to United States military officials that the most appropriate place to establish their regional headquarters was in the small bayside village. The population subsequently grew with the arrival of a shipload of Mormons in 1846. By 1848 Yerba Buena had been renamed San Francisco and the population had reached about 900.

Sam Brannan's pitch caused astounding demographic changes. Initially, the town quite literally emptied; upwards of three-quarters of the residents grabbed picks and axes and scurried to Coloma, prompting some to suggest that the place looked like a ghost town. The situation did not last long, because the subsequent half-dozen years saw dramatic population increases the likes of which history had rarely witnessed: 25,000 in 1849; 34,000 in 1852; and 50,000 by 1856. By mid-decade the city's future was secure.

If not for a fortuitous matter of topography, it's doubtful that San Francisco's boom would have happened. The rush caught the town unprepared, without the many wharves necessary for ships to dock and unload their goods, or enough warehouses to store the supplies. As a result, throughout the early months of 1849, incoming merchants and miners alike bypassed the town and went directly upriver to Sacramento. Business interests in San Francisco were quick to see the financial possibilities of the rush, so it didn't take them long to build the necessary accommodations. But it was a change in the nature of shipping that

Nothing illustrates the enormous scale of the California Gold Rush as well as the meteoric population explosion of San Francisco. A sleepy frontier village of no more than 900 souls in early 1848, by the end of 1849 it was a rough and rowdy home to over 25,000 fortune seekers. Seven years later the population doubled to 50,000. San Francisco was the first boomtown of the American West; the rough urban landscape would be replicated in numerous other frontier towns that sprang up throughout the West in the latter half of the 19th century. Crime abounded in the saloons, brothels and gambling houses that proliferated during San Francisco's early years. But unlike so many of the West's boomtowns, San Francisco outlasted the precarious economic circumstances of its origins and acquired the trappings of an established municipality.

sent San Francisco's star soaring. As increasing numbers of miners demanded passage, larger ships were required. Entrepreneurs from other cities knew that the biggest profits were to be made by those who got their goods west fastest. Slower vessels often arrived to glutted markets, causing anticipated fortunes to melt away. To cut travel times, steamers and clippers were contracted. These vessels could not navigate inland waterways and they had to call into port at San Francisco.

With the construction boom, San Francisco underwent an incredible transformation, none of it characterized by planning or orderliness. The city was built from the harbor fronts up the sloping hills and, during the rush, lowland was always more valuable than higher ground. To extend the lowlands, hills were scraped level and the resulting gravel was emptied into the coves that eventually created new lots in the harbor. Without government planning, folks built wherever they wanted. The only zoning was the result of the outcry from local citizens, who were quick to object to stinky slaughterhouses and acrid soap factories in their backyards. Most wharves were also privately built, and those who constructed them found that they had struck their own gold mines. Since all goods came up from the harbor, they had to be unloaded on the docks. The merchants were smiling. They all made money, giving credence to the old cliché that the surest way to make a fortune during a gold rush was to be a store owner.

Not many months had passed before the shoreline was a jumble of incongruous wharves. The receding landscape was a compact jumble of hastily constructed, multi-sized warehouses, a varied collection of manufacturing establishments, retail outlets offering all sorts of goods, and residential dwellings from lowbrow canvas covered shanties to more respectable multi-story frame houses.

The disarray evident on land was reflected in the conditions of the harbor. On any given day the rolling forest of masts was packed so tightly that a determined sailor who found his vessel on the outer margins of the harbor could easily make it ashore by hopping from one ship to another. He'd have to exercise some caution, however, because there were plenty of scuttled vessels left to rot after they'd limped into port. Those that had not yet been rendered skeletal by builders scavenging for wood or had succumbed to rats, woodworm and salt were used as seaside quarters with the least comfortable serving as warehouses or prisons. Perhaps such a choice for a residence is more understandable when considering the cost of land. Real estate speculation had sent prices soaring. Waterfront lots that sold for tens of dollars in the mid-1840s were commanding thousands, and sometimes tens of thousands, of dollars a handful of years later. If a fellow wanted to rent a space for an office, he was looking at as much as $1000. As a contemporary described it, San Francisco "seemed to have accomplished in a day the growth of half a century."

It wasn't only the physical structure of San Francisco that was transformed; the face of the town was also literally recast. Almost overnight, San Francisco became a cosmopolitan city long before the description was in vogue. French berets and Mexican sombreros bobbed along busy city streets. The long braided hair of Chinese and Natives swayed without restraint. The dark flannel shirts of the Americans contrasted poorly with the colorful serapes of the Chileans. And the languages that greeted the attentive ear! German, Kanakas, Russian, and as many English accents as there were colonies of Britain. Throughout the gold rush years, over 50% of the city's population were non-Americans. San Francisco offered the possibility of prosperity for all and it drew immigrants without discrimination, although there would be enough of that when folks arrived at the diggings.

Those forty-niners destined for the gold fields spent as little time in San Francisco as possible. Usually they were on the next boat bound for Sacramento and, at most, that meant a couple of nights in a hotel, a boarding house or often a tent. Even as short a stay as that was likely to send many new arrivals into a state of shock. One look at the price of goods in San Francisco was enough to deter most folks from calling it home. Deals were done in dollars with the lowly copper being, as one fellow put it, "a thing of antiquarian interest." It's enough to know that eggs were $1 each; apparently a grocer didn't need to steal from a giant to have his own bag full of gold. Those who failed to buy their mining equipment in the east were crestfallen when they discovered that hip-waders were $100 a pair and that shovels and picks would set them back $10 apiece. Sam Brannan had bought all the mining pans he could before the rush for $.20 each and was later selling them for as much as $16. "California prices" quickly became a derisive term for overpriced goods. Of course, wages were high so as to meet the inflated state of affairs; a laborer made about twice as much as he could expect to earn in the east. And, if a forty-niner was lucky and had something to sell, then he could significantly cut his losses. One fortunate fellow discovered that his eastern newspapers, months out of date, were useful for more than wrapping fish. He sold them at a 4000% mark-up.

If "California prices" were insufficient to hurry a God-fearing miner to the diggings, San Francisco had plenty more to offer that could prove equally effective. There were folks swept along in the rush who saw their fortunes as more certain if they brandished a revolver or knife rather than swung a pick, especially after they'd tried a spell in the diggings. New arrivals were generally considered to be unknown quantities and were treated with suspicion, as was well illustrated in a contemporary song,

> *What was your name in the States?*
> *Was it Thompson, or Johnson or Bates?*
> *Did you murder your wife and flee for your life?*
> *Say, what was your name in the States?*

There were plenty of individual operators, usually known collectively as Sydney Ducks (a reference to the Australian penal colonies). The French generally had a bad name because of the Lottery of the Golden Ingot. Louis Napoleon, the French president, dreamed up the idea and promoted it as a scheme to allow his poorer citizens access to fortune. In reality, he was ridding the country of his enemies and the criminal element, but few complained when given a crack at California gold. There were also terrifying gangs, the most prominent of which was named the Hounds. It terrorized San Francisco for the better part of two years, prompting one prominent resident, a clergyman, to reveal, "Most of our citizens, if not all, go armed." The Hounds were not brought to their knees until the community established its own volunteer police force in the summer of 1851. Even then, it was rare day when there was no news of an assault, a robbery, a murder, or worse, a fire.

Fires were the greatest threat to the city's property and its residents' safety. From the outer reaches of the bay to the gradually sloping hillsides of the cove, San Francisco was a wooden city. From ships, wharves, planked roads, buildings and houses, there was little that didn't use wood as an integral part of its structure. What wasn't wood, was canvas or cloth. During the hot, dry, windy summer, sparks from chimneys, or a careless encounter with a lamp often had expensive and deadly consequences. The greatest concern was not accidents, but intentionally set fires. When waterfront property was at a premium during the gold rush, cutthroat businessmen employed firebugs as a way to remove competition and clear land for sale. In the first half of the 1850s, estimates indicate that nearly one-third of the city's hundreds of fires were intentionally set. Some caused staggering damage. Two in the spring and summer of 1851 destroyed more than 30 city blocks, at least a third of the city.

Tiring of the lawlessness and incendiarism, some city merchants banded together to form the Vigilance Committee, though some suggested that the solution was almost as bad as the problem. Numbering about 500 members, the Committee's nightly patrols ran roughshod over the rights of residents, although only those who fell victim to their activities did much complaining. Even then, fear usually muted their words. Often with little more justification than rumor, they searched without warrants and deported the suspicious. They also had a habit of hanging those they considered guilty. The Vigilance Committee wasn't really much more than mob rule run amuck. They were urged on in their efforts by San Franciscans who were, for the most part, just as supportive of harsh punishment as were the members of the Committee. There were good reasons for a fellow to show the city his back as soon as he could.

By the mid-1850s San Francisco had come of age. It provided all of the amenities one expected to find in a developed eastern city and plenty of the adventure and recklessness most of those cities were happy to be without. The cultured could spend their time at operas and ballets.

One might say that hard times called for hard measures. The first Vigilance Committee, established in 1851, was a response to the lawlessness on San Francisco, streets during the first years of the gold rush. When the local authorities proved unable to control the rampant crime, a group of over 500 San Francisco shopkeepers, businessmen and bankers banded together to take the law into their own hands. The Vigilance Committee's idea of justice was neither impassive nor orderly—very often was not even fair—but it got the job done. Rampaging gangs of vigilantes patrolling the streets at night exercised judgment as harsh as it was swift and rarely were they concerned about individual rights. By the end of its first year, the committee had deported 20 criminals, whipped one transgressor and hanged four others. There was no question that crime dropped markedly, and the committee disbanded after its first year. A second Vigilance Committee was formed in 1856, but didn't last as long as the first, breaking up after three months.

Desires could be sated at gambling houses, race tracks and brothels, none of which were wanting in number. Miners had money to spend and were mostly inclined to spend it. Houses of ill repute were particularly popular, because one of the features of the rush of '49 was that it drew few women. For the most part, only Mexican miners brought their wives with them, and it showed in the city's demography. In the early 1850s, the ratio of men to women was easily 10 to 1 (in the diggings, it was even higher). A few years later a more equitable balance was achieved and settled family life increasingly became the norm. As the gold rush petered out, San Francisco was able to build on its regional dominance and become the prominent metropolis for much of California and the territory to the north.

In a canyon west of Sacramento, there stood a miner forty-niner. He was one of the youngest of 50 or so in a small mining camp still too new to have a name. He heaved a great sigh and thought some more about the task at hand. Better balance, that's what he needed. He lifted one leg and placed his foot on a protruding rock. He steeled his eyes and clenched his jaw tight. Any miner who might have looked his way would have noted his determination but little would have been made of it. Any forty-niner worth his salt threw everything into his work. The green miner took one last look at the pointed contraption in his hand, raised it high in the air and then, with a sudden swinging motion, he came down with it, driving it right through his finger! He held his hand up and stared at it. The blood seeped out of the wound until it was dripping down his wrist. The sight and the pain were too much for him and the canyon echoed with his howl of pain. Miners knew the difference between a cry of joy and one of agony, so they hurried to see what the problem was. A late arrival grabbed the shoulder of one of the men who surrounded the injured miner.

"Wha, what happened," he panted.

"Young Jim here just figured out that he got some learning to do if he wants to sew!" came the chuckling reply.

Mining was hard work and many a fellow was called to do what he had never done before. Occasionally the results were hilarious, especially when the miners had to perform the domestic chores that were traditionally performed by women, of whom there were precious few in the mines. And, of course, men came to the gold fields to mine. Any activity (save recreational ones) that kept them from their claims was ignored or, if necessary, poorly done with haste and without enthusiasm.

Clothing always presented the greatest challenge. Working in piles of jagged rocks was hard on outerwear, especially pants, and eventually all miners had to learn to sew. To the chagrin of many, they also had to wash their clothes, a fact that made dark apparel fashionable. Sooner or later, the buildup of dirt and sweat was such that a miner couldn't even stand to work alone and he was forced to the river to clean his clothes. Even then, some preferred a less demanding procedure as described by one miner. "Have two shirts. Wear one until it is dirty. Hang on a limb, exposed to wind, rain, and sun. Put on second shirt. Wear until dirty. Then change to the clean one." The undertaking didn't result in the freshest of shirts, but it was efficient.

Aside from the issues involving clothes, the greatest challenge was cooking. There was never much variety in mining camp grub; it usually consisted of salted or jerked meat, hardtack, beans and flour, all foods that kept well enough, but that didn't arouse much excitement at meal time.

Though there was often the possibility of fresh game or fish, most miners preferred to work their mines and were willing to do without. Similarly, they didn't like to spend a lot of time cooking. So, meal preparation was frequently rushed and done by exhausted miners in the flickering light of a candle or fire when it was too late to mine. Those conditions weren't always well suited to cooking, as one miner discovered when he mixed together a helping of hardtack and salt pork. The next morning he discovered that insects had infested his supply of salt pork and he had to face the queasy reality that he had gorged himself on a meal of bugs. Other times, the food simply went bad (not uncommon in the heat of a California summer), but most miners weren't aware of that until their stomachs started churning a few hours after they'd eaten.

Men often got sick in the mining camps. The food itself (or lack of it) sometimes caused diseases like dysentery or scurvy but, more often, poor nutrition merely lowered their physical immunity to sickness. There were plenty of places a bug or a microbe could call home in the diggings. Little effort was devoted to matters of sanitation and the result was predictable; the camps were breeding grounds for disease. Cholera and typhoid became common enemies. Working in stale, dank conditions also contributed to illness, and there were more than a few miners who felt the aches and pains of rheumatism and tuberculosis. Sometimes disease was beyond anyone's control; there was little that could be done about the ague and other fevers caused by an insect's sting. Most mining camps had a store of medicines lying about, including laudanum, quinine, and calomel. However, these were self-administered in a most unscientific manner, that is, miners guessed as to the appropriate cure and the dosage. Occasionally a doctor showed up, but often the sick would have been better off had he not. Most of the self-professed physicians were quacks, pushing some concoction of snake oil on unsuspecting miners. The simple fact was that mining was a dangerous business with many deaths, and not all of them from mining accidents.

Neither could a sick miner find solace recuperating in his living quarters. In most camps even the healthy didn't enjoy turning in. Sometimes a forty-niner might get his wagon to the diggings; these lucky ones actually had something substantial over their heads. Most mining camps eventually had their share of wooden cabins, either of log or split timber, but they were characteristic of more mature camps and, in any case, rarely built before the end of the mining season. The cabins were more serviceable than comfortable, usually consisting of bunk-beds, barrels for tables, chairs and some sort of fireplace. In the early days of the rush, most miners lived in tents. Given that it was frequent for a tent to have more sleeping in it than its designers intended, some took to slumbering under the stars. Such a bed avoided the indecent smells of over-crowded canvas quarters and it also kept the lice at bay.

A mining camp that wasn't infested with vermin didn't exist. Many mindless minutes were spent inspecting clothes for the tiny pests. For the most part, they were simply endured, though miners, always an optimistic lot, occasionally made the best of an irritating situation. They engaged in lice races. The miners took the contests seriously and substantial wagers were often placed on would-be thoroughbreds. A revealing story involves the champion of an Irish forty-niner. His louse raced to an easy victory. "It's hardly surprising," proclaimed the owner. "This louse has the blood of Irish royalty." An onlooker who had backed the wrong critter approached the victor. "Might I borrow your louse?" he inquired. Then, with a straight face, he explained, "I want to improve the breed of my own stock."

There were other forms of recreation in the mining camps and most of them enjoyed their greatest popularity in the winter, when there was plenty of time and few exhausted miners. Nothing topped cards, and whether the games were wagered upon or not, many an hour was whiled away behind poker faces. If a fellow had a book, he might well be the most popular man in the camp, especially if he was inclined to read a few pages aloud each night. And those willing to share a letter from home briefly enjoyed the status of royalty. There were always some musical instruments that men played around the campfire late at night. Occasionally miners craved some excitement and so the instrumentalist found himself fiddling or blowing on his harmonica at an impromptu ball. Since women were rare in mining camps, when it came to the dancing, some miners (identified by their handkerchief) became the belles of the ball. They'd never been so popular!

Recreation wasn't only a way to pass the time or relieve boredom; it was as valuable in relieving loneliness. Men pined for home, especially mothers and girlfriends. One fellow missed the sight of women so much that he actually walked 30 miles just to look at one. It's likely his story wasn't unique. But the best evidence of homesickness was found in the wistful songs that became popular around the crackling fires. Songs like *Do They Miss Me at Home* and *Then Hurrah for Home!* speak for themselves. Occasionally they even reveal the lesser known aspects of camp life. Such is the case with *I'm Sad and Lonely Here*, which deplores the everyday violence of the mining camps, "I dare not speak, for fear some fighting man will pound me." But perhaps the longing sentiment for home is best summed up in *The Miner's Dream*, a song that might have been written by the great poet Blake.

The miner when he goes to sleep, soon begins to snore;
Dreams about his friends at home, whom he may see no
  more;
A lovely wife or sister dear he may have left behind;
Perhaps a father old and gray, a mother good and kind.
Now will you, say you will, listen while I sing
A song that's called the miner's dream? 'Twill joy and comfort
  bring.
His boyhood years return again, his heart is filled with
  joy—
Is rolling hoops or playing ball as when he was a boy.
'Tis winter time—he's skating now, of which he was so
  fond;
'Tis summer now—he's swimming in the old familiar pond.
His boyhood days are past and gone, for now he is a man—
Is going to California to try the pick and pan;
Bright visions now of happiness are dancing o'er his
  mind;
Disturb him not, but let him dream so long as he's inclined.
His mind is home among the fields of wheat and yellow
  corn—
Sits down beneath an apple tree, all shady in the morn—
But morning comes—and at his door a neighbor gently
  knocks;
He wakes, and finds himself among the hills and rocks.

In the mining camps, all that glittered truly was not gold and men dreamed of more than fortune.

Miners were there to strike it rich, and that meant that as many daylight minutes as possible were devoted to pick and pan, shaft and river. First they had to learn how to mine. Most forty-niners arrived ignorant about how to get gold from stream or rock. They learned quickly that the "California Gold Grease" sold in the east was of little value. One glance at a mountain made it clear that a fellow wasn't going get rich rolling down the mountainside, no matter how much of the grease he smeared over his body. And the psychics that some brought out to help pinpoint the deposits proved to be little help. Unless one was lucky enough to strike it rich with one swing of the pick (and there were such stories, like Rich Bar), mining demanded physical labor and long hours.

Most miners tried the pan, though as often as not, it was a frying pan with remnants of pancake still attached rather than one designed for prospecting. A good day would see a miner work through 50 pans full of water and gravel, but bending over the stream was slow backbreaking work. They were always searching for more efficient means of recovering the gold—the rocker came out of that search—but most forty-niners spent their longest hours in the mines. Mining there was not more challenging than swinging a pick and filling a bucket with the lose gravel. It did, however, come with its dangers. The pine boxes filled with those who had died of illness were supplemented by those who died in cave-ins, if their bodies were ever recovered. The simple facts were that mining was not an exciting activity and there was little to be said for it, beyond occasional ingenuity, other than the time it demanded and the pain it caused. It explains why so little effort was devoted to the other aspects of camp life. As one surprised miner informed his eastern friends, "I can assure you that gold digging makes a man sleep well."

The sun was just peaking above the Sierra Nevada; the morning sky was blood red. Two shabbily dressed and tuckered miners were following an easy route out of the Mother Lode, the foothills where the rich southern

For the first prospectors who arrived at ore-rich sites in northern California, gold mining was a simple, if physically demanding, procedure. These miners could often dig gold out of rock surfaces with a hunting knife or a pickaxe. But it would not be long before all the easily visible nuggets were gone, and miners would then have to turn to placer mining the sand from the rivers and creeks. Gold panners were by far the most common miners found on the riverbanks, men hunched over their swirling pans, eyes hungry for the flash of gold from the gravel in the bottom of their bowls. It was backbreaking, monotonous work in a potentially explosive environment. If a waterway was rich in gold, it often grew overpopulated with panners, and then violence erupted as men were convinced that they were being robbed of ever-decreasing gold yields.

California gold fields were located. The men had spent five tiring months coyoting, sinking holes into bedrock and digging out side tunnels to the gold-bearing gravel. The process was referred to as coyoting because coyotes were known to live in crevasses and holes in the hillsides. The pair of miners was headed to Stockton, by way of the San Joaquin River. Ropes, tied to their saddle horns, were attached to a couple of pack mules. The animals had it pretty easy on this trip. The men were low on supplies and needed to fill their stores. When they got to Stockton, however, they planned on doing more than loading up on dry goods. The miners had done well and between them they had $20,000 worth of gold. They'd be buying evening wear and drinking champagne from slippers. By God, Stockton wouldn't know what hit it!

The miners weren't out of the foothills when a shot rang out and kicked up a puff of dust on the trail before them. They fumbled for their rifles but before they could get them out, they were suddenly surrounded by a band of men and looking down the barrels of a dozen rifles. One member of the gang spoke in a thick Mexican accent, ordering the miners to give up their gold.

"It is better for you to save us the trouble of looking," he added.

The miners hesitated. They stared at the man who'd addressed them and concluded from his appearance that he was the leader of the outlaws. A gold-laced jacket concealed most of his bright red shirt and silver disks dangled from his pantaloons. Large silver rowels could be seen at the heel of his well buffed knee-high leather boots and a fringed blue cloak rested on his back. Although his facial features were concealed in the shadow thrown by his plumed sombrero, the miners could see the glowing red tip of a cigar. The outlaw raised his chin. The miners watched mesmerized as he exhaled a cloud of smoke and the gray fumes slowly crept along the underside of the rim of his sombrero, finally escaping and slipping skyward. They gazed too long, so the man gave the order to shoot them. Joaquin Murieta spurred his horse over towards the dead men. He pushed one of the bodies off its horse and it fell to the ground with a thud. Murieta and his men found the gold and rode off, leaving the miners as food for the carrion eaters.

There were many outlaws in California during the rush, but few gained the infamy of Joaquin Murieta. Perhaps it had something to do with the fact that he was a Mexican. American forty-niners did not like Mexicans. They were recent enemies; they weren't white; they were considered interlopers, believed to be in California simply to mine gold and return to Mexico; and they were too numerous. Along with the Chinese and the Natives, the Mexicans were at the bottom of the barrel as far as most Americans were concerned. The attitude goes far towards explaining why men like Murieta emerged as heroes among their people. To many white miners he was a thief and a murderer, but to Mexicans he was El Patrio, a military champion, striking a blow against American domination and racism. His story only galvanized his Mexican support.

Joaquin Murieta, the outlaw, was born in the violence that characterized much of the California Gold Rush. Murieta was drawn to California, like most forty-niners, with visions of gold glittering in his eyes. He staked a claim and started to dig. One day some miners stumbled upon his claim and demanded that he give it to them. When he resisted, they beat him, tied him and forced him to watch as they violated his young mistress, Rosita Feliz. He escaped with his life, but his troubles weren't over. He was subsequently caught riding an allegedly stolen horse. His pleas of innocence fell on deaf ears. He was whipped while his brother, who rode with him, was hanged. Murieta then became an outlaw and the first to die were those who had slipped the noose around his brother's neck.

It didn't stop there. Murieta organized a gang of cutthroats who were scary even in the light of day. At the height of his power the Mexican outlaw commanded more than 100 men, who were divided into several companies, but who always answered to him. They rode through the gold country with abandon and soon there wasn't a county in all of California that the Murieta gang had not struck. Although there is no accurate record of the amount Murieta stole, it's reasonable to believe that it was well into the hundreds of thousands of dollars. He didn't spend all of it. Murieta was known for his a generosity to his needy countrymen, earning him the name "Robin Hood of El Dorado" in some quarters. And even today rumors persist that parts of his treasure remain buried throughout the state.

Eventually Murieta's marauding was such that California lawmakers could no longer ignore him. The authorities posted rewards and the California legislature passed a bill authorizing the creation of an armed contingent whose sole purpose was to bring Murieta to justice. The company was named the California Rangers and its 20 men were under the command of Captain Harry Love. Murieta's cunning almost made his escape from even those battle-tested pursuers. When he became so well known as to be identified by sight, Murieta took to wearing elaborate costumes, and he was as likely to strike, or escape, dressed as a priest or a woman as he was a bandit. Captain Love actually stumbled upon Murieta and his gang, but the outlaw was disguised as a ragged peon, so Love was unaware of his luck until a member of the California Rangers who knew Murieta pointed him out. Though he tried to flee, Murieta was shot.

Murieta terrorized California for three years starting in the summer of 1850, but his fame was such that his flame burned long after his death. An entrepreneur (California was never in want of such men) managed to acquire Murieta's head and he made a good living putting it on traveling display. Other outlaws also enjoyed infamous reputations during the rush of '49. Three Finger Jack Garcia, who rode with Murieta, and Black Bart, who terrorized stage coaches, were two of the better known outlaws. But the majority of those who committed crimes were usually without reputations, mostly because they didn't live long enough to enjoy the fruits of their felonious activities.

There's a story told about a forty-niner whose actions apparently crossed the line of miner acceptability. The exact nature of the alleged crime is lost to history, but it was serious enough that the dozen men brought together to listen to the evidence knew that a guilty verdict would result in a hanging. After some consideration, the jury found the man not guilty. Upon hearing the decision, an onlooker tipped the rim of his felt hat low so that his eyes were lost in its shadow, and with a slow measured pace, he approached the jury.

"You boys best do better than that," he muttered.

The jury reconsidered their decision and quickly reconvened. Within minutes it returned a revised verdict of guilty.

"That's the right call," shouted someone in the crowd. "We hung the bastard an hour ago."

Such stories of crime and frontier justice in California's mining camps were almost as legion as those of gold strikes. In the early months of the rush, crime was uncommon, because honest miners respected the toil and profits of others. Malfeasance was inevitable and soon incidents of theft, be it of equipment, gold or horses, were daily

WILL BE EXHIBITED
FOR ONE DAY ONLY!

AT THE STOCKTON HOUSE!
THIS DAY, AUG. 12, FROM 9 A. M., UNTIL 6, P. M.

THE HEAD
Of the renowned Bandit!

JOAQUIN!
AND THE
HAND OF THREE FINGERED JACK!
THE NOTORIOUS ROBBER AND MURDERER.

Murderous thief or Mexican folk hero, Joaquin Murieta (1830–1853 or 1878?) was a six-gun marauder who made his living robbing California gold miners of their hard-gotten gains and sharing it among the more unfortunate of his countrymen. Countless legends surround the bandit, enshrouding his life in as much mystery as his death. Harry Love and his California Rangers were rewarded $1000 after they sent the head of a man who was supposedly Murieta to the governor. The head, along with what was said to be the hand of his partner, Three-Fingered Jack, was pickled, stuffed into a whiskey jar and circulated through the mining camps and towns in the territory. Great crowds gathered to see the grisly trophies, but it was not long before rumors began to spread that the head was not that of the famous bandit, that Murieta himself was alive and well in Mexico. To this day, it has never been affirmed who the "Robin Hood of the El Dorado" really was, if it actually was his head that was displayed through California.

occurrences and it was a rare week that wasn't witness to a murder. Since any official lawful presence seldom arrived in mining camps until the early 1850s, most forty-niners took to enforcing their own brand of justice, and retribution figured prominently. An onlooker would have needed to witness a bushel full of proceedings before he saw either the wisdom of Solomon or the mercy of Christ. In criminal matters innocence became a relative term, relative, that is, to how likely a verdict was to deter future criminals from the wayward path. Given the importance of deterrence, a harsher punishment meant a better one. The enforcement of justice also had an interesting effect on the miners. For the most part, those who mucked in muddy ditches and cold shafts were a fiercely independent lot. But when a crime was brought to their attention, rugged individualism rapidly gave way to a mob mentality and the accused often found themselves subject to judgement by those for whom any inkling of rationality was a dim spark at best.

It wouldn't be right to paint too dark a picture of law enforcement in the gold fields during the rush of '49. There were mining camps that were decent enough places, administered by rules, known as the claim laws, that were open to debate and voting. Most claim laws focused on the mining itself, for example, ownership, use and water rights. The rules weren't always for the good of all; occasionally the racist attitudes of the time surfaced in the form of directives that forbade Asians and South-Sea Islanders from mining in a district. If there were disputes that could not be resolved by the claims laws, the alcalde (claims officer) stepped in. Often the alcalde—a term that was a throwback to the old days of Spanish rule—was the only elected official in a mining camp, but his authority was mostly in civil matters.

When it came to criminal issues, law enforcement was much more of an ad hoc affair, a scaled-down version of the acceptable norms of jurisprudence, one without the frills. The advocates of the law, like lawyers and judges, were generally seen as impediments to justice, and therefore best done without. Although theirs was a swifter form of justice, it was rarely an improvement. If a crime was committed, 6 or 12 men (or sometimes the whole camp) would gather together, listen to the testimony and give their verdict. The accused fell into one of three categories: the guilty, the innocent and those no one could be sure about. The only thing that could be said for certain about each category of defendant was that they shared about the same chance of being convicted. The attitude was well summed up in a contemporary story detailing the advice given to one jury: "Take your time, gentlemen, but remember, we want this room so's we can lay out the corpse in it."

It was uncommon indeed when justice did not move with lighting speed. Since miners had better things to do than build jails and guard them, punishment was usually the corporal kind, such as hanging, whipping, branding, or ear-cropping or else banishment. The most popular seemed to be hanging, as suggested by reports that described the tree-lined trails of California's interior as "tasseled with the carcasses of the wicked." There were plenty of cases where the charge was read and the body was swinging from the branches of the nearest oak in a time span measured in minutes. Under such circumstances there were tragic occasions when the hanging rope was stretched tight only to discover that the deadweight in the noose was innocent. Such news was usually met with a shrug of indifference; it was an acceptable price for maintaining order.

The California Gold Rush was at its peak between 1849 and 1852. There was still plenty of gold mined after 1852, but the crazy boom-time years became only a memory. San Francisco was maturing into a regional metropolis. In the diggings, law and order were breaking out, prompting one miner to lament that there were "no fights, no murders, no rapes, no robberies to amuse us." Most significantly, the nature of mining changed. The easy pickings had been cleaned, true prospectors had moved farther west to scratch their itches, and most operations were expensive, labor-intensive hydraulic mining and hard rock mining in quartz veins. By 1853 California was less a place of adventure than it was of big business and bigger machines.

But, oh, what a four years! The average gold production during the period was some 2,700,000 troy ounces. In 1852 alone, forty-niners panned and pried just a shade less than 4,000,000 ounces. And there were many to do the work. In 1849, newcomers numbered 80,000, with another 90,000 coming in the following year. In the handful of years that witnessed the gold rush frenzy, over 250,000 came looking for gold. Many of them stayed and by the end of 1852, the population of California was pegged at 223,856, quite an increase from the few thousand non-Native and non-Mexicans who called the place home before the summer of '48.

A small percentage of the great number who came actually struck it rich. There were stories of lumps of gold that weighed more than 20 pounds and many of them were true. But most miners didn't do more than make a living, though it was a good one. The average was $25 a day and, even with the inflation of gold rush supplies, that still looked pretty good compared to the $2 to $3 a day a skilled worker might expect to make in California ($1 a day back east). But it wasn't the prospect of $25 a day that drew men to America's west coast; it was the dream of the big strike. Who could hear of the 28-pound lump of gold unearthed at Sonora, the 273 pounds taken from the Feather River in 7 weeks, or assertions that the riverbeds were paved with gold and not get giddy? Even the subdued appraisal of Colonel Mason proved magnetic. "No capital is required to obtain this gold, as the laboring man wants nothing but his pick and shovel and tin pan with which to dig and wash the gravel." But it wasn't all gravy, as one forty-niner soon discovered. "It is six weeks since I reached the mines, and they have been rendered memorable by the hardest work I have ever undergone."

Americans formed a large majority of those who flocked to the gold fields; tradition has it, in fact, that everyone on the east coast had a relative or friend of a friend of a friend digging in California. They quickly became part of the great international assembly that was California in the early 1850s. Probably no country saw a larger proportion of its residents leave than Australia—1 in 50! Many, save the Mexicans and the local Natives, were foreigners who traveled great distances to join in the rush. Those who differed most from the Americans in color and culture faced discrimination and were told in no uncertain terms that they were unwelcome. The attitude is well summed up in the lines of a popular contemporary song, *The National Miner*,

> Here we're working like a swarm of bees
> Scarcely making enough to live,
> And two hundred thousand Chinese
> Are taking home the gold we ought to have.

The California Gold Rush gradually receded into history after the mid-1850s, but its effects on the American West were enormous, setting the foundation for settlement in the "Golden State" and galvanizing the collective imagination with the tonic of opportunity. The romantic image of the rugged gold miner working alpine streams for nuggets resonated with an American population that was tirelessly in the pursuit of happiness and fortune. The miner became an American icon; a figure that suggested anything was possible with hard work, tenacity and a little luck. In reality, the life of the miner was hard, dangerous and likely to produce heartbreak and disappointment. Idyllic depictions of gold miners such as the above Currier and Ives lithograph abounded in the 19th century, and the West became a byword of opportunity for following generations.

Mostly, however, miners got along well enough. They had to, since they were thrown together in close quarters for extended periods of time. Those who caused problems were dealt with harshly, but that was a reflection of the tough, physical life the forty-niners endured.

When it was all said and done, the California Gold Rush was about much more than numbers and production. There were gold rushes before and after 1849 and some had bigger strikes than those discovered in America's far west. What happened in California changed a nation. It was the spark that caught the imagination of Americans and etched the west as a place of adventure in the popular consciousness. It allowed Horace Greely, editor of the influential *New York Tribune*, to advise ambitious men to go west. Many did and the future of the United States as a transcontinental nation was thereby solidified. For what it did for the country, the rush of '49 holds a special place in the American memory, as is evident in the songs that even today roll off the tongue with ease.

> *In a cavern, in a canyon, excavating for a mine,*
> *Dwelt a miner forty-niner and his daughter Clementine.*
> *Oh my darling, oh my darling, oh my darling Clementine,*
> *You are lost and gone forever, dreadful sorry, Clementine.*

The Fraser River
(1858)

JAMES DOUGLAS RAISED HIS HEAD from the stack of documents that lay before him and rubbed his tired eyes. He rose, stepped from behind his desk, walked to the window, and rested his hands on the sill. It was April 1858 and the unseasonably dry, mild winter had long since given way to a most promising spring. Douglas threw open the window and breathed deeply of the salty air, sweetened with a hint of spring's first fragrances. Was that a hint of heather in the breeze? He allowed himself to daydream briefly. If any man was justified in taking a moment from the rigors of work, surely it was Douglas. He served as both the regional Chief Factor of the Hudson's Bay Company and the colonial Governor of Vancouver Island. He had filled the commercial position for nearly 20 years and had enjoyed the confidence of the Queen for the previous seven years. But the challenge to balance the sometimes competing interests of business and empire had been tiring.

A rap on the door brought Douglas back to the present. His jobs were about to get downright exhausting.

"Come," called Douglas.

His assistant slipped through the door.

"Sir, the *Commodore* is entering the harbor. You wanted to be informed before the passengers disembark."

"Right," replied Douglas as he rubbed the back of his neck. "Prepare to walk down there. I'll greet them as they leave the dock. Ensure that there are copies of the Gold Proclamation for posting. Not that I expect many of them will be literate," he added.

It wasn't every day that Douglas made his way to the waterfront to meet the colony's newcomers. But the passengers on board the *Commodore* weren't just any newcomers. They were the first large scale arrival of gold prospectors bound for the mouth of the Fraser River and points beyond—Fort Yale (just south of the rapids), the Couteau Mines (just south of the fork of the Fraser and Thompson Rivers) and many other places that had not yet been named. They came north on the basis of little more than rumor. Even though Douglas' own reports couldn't provide reliable information on the amount of gold in the region, there was concrete evidence of its existence. That the prospectors were aware of it was as much Douglas' responsibility as anyone's. In February, he had dispatched a Hudson's Bay Company ship to San Francisco, its cargo of 800 ounces of gold bound for the local mint. Such news did not remain a secret for long.

Coming from San Francisco, the newcomers packed a mean reputation. All reports suggested that they were among the worst elements of the southern city. Douglas wasn't even sure that the passengers realized that they'd be stepping ashore onto British territory. Well, by God, they'd learn fast enough. Before these men caught sight of a glimmer of gold, they'd know that Britain owned this stretch of the Pacific Northwest and that, like all things British, it had

Dubbed the "Father of British Columbia," James Douglas (1803–1877) was the most powerful man in the Pacific Northwest in 1858, holding the titles of Chief Factor of the Hudson's Bay Company in Fort Victoria and the governor of Vancouver Island. He was an adept, if stern, patriarch, whose lifetime of work with the HBC instilled in him an amazing capacity for administration. But Douglas was as courageous as he was organized, and there were few others who would have been better able to manage the chaotic rush of gold prospectors to his erstwhile peaceful colony in 1858. Every governing decision that he made during this time was aimed at keeping the area under British rule. His two-pronged strategy of subjecting the miners to British law while providing an easy transportation route to the gold fields of the interior worked, and the far-flung British colony survived the sustained gold rushes throughout the late 1850s and 1860s.

rules to be obeyed. If the Governor's presence didn't ensure that, the Gold Proclamation would. The document set the tone and the general framework within which prospecting could be undertaken. In part, it read that "all persons who shall take from any lands within the said districts any gold, metal or ore containing gold, or who shall dig for and disturb the soil in search of gold, metal or ore without having been duly authorized in that behalf by Her Majesty's Colonial Government, will be prosecuted, both criminally and civilly, as the law allows." To avoid prosecution, prospectors had to obtain a mining license. And nobody got a license unless he was willing to undertake an oath of allegiance to the British Crown.

It takes little imagination to envision the surprise on the faces of the prospectors, both cynical old veterans and wide-eyed tenderfeet, as they were greeted by Douglas's regulations for prospecting. Not all the newcomers were impressed by what they heard. Gold rush enthusiasm quickly turned to disbelief and eventually anger. The waterfront soon screeched with howls of protest.

"Rules? Rules!? There ain't no rules in a gold rush!"

"It's e'ery man fer hisself!"

"Just point us in the right direction, guv'nor. We can figure the rest. You don't worry yourself about us."

As the agitated men fed off each other, the scene soon turned ugly.

"Suppose we came and squatted," said one in a tone that clearly suggested it was meant to be more of a statement than a question.

"You would be turned off," replied a calm Douglas.

"And if several hundred of us were prepared to resist, then what would you do?" continued the defiant prospector.

"We should cut them into mincemeat, sir, we should cut them into mincemeat." It wasn't the threat, but the matter-of-fact way in which the message was delivered that silenced the aggressive miner. Apparently the Governor meant business.

And he did. For years, Douglas had anticipated the arrival of the miners and he had spent long hours reflecting on the most effective strategy to maintain order and to uphold the rule of law. The preparations he made did much to ensure that the Fraser Gold Rush would be mostly a peaceful enterprise, much different that those experienced by its southern neighbor.

༜

The early history of the Pacific Northwest gave little hint that there would be much difference between the gold rushes in British territory and on American soil. Like California, a European presence on what would become known as Vancouver Island and the adjacent mainland was relatively late in coming. Though Frances Drake had apparently sailed through the coastal region in 1579 and there were rumors of Russian fur traders active there a century later, it wasn't until the late 1700s that Europeans officially began trading in the Pacific Northwest. The three major players were the Russians, the Americans and the British. Each had trading enterprises in operation—the Russian American Company, the American Fur Company, and the Hudson's Bay Company, respectively. They were there for sound economic reasons. The region was rich in fur-bearing animals, especially the sea otter, whose pelt was in high demand in the Oriental marketplace. And just as important was the willingness of the local Natives to trade.

Throughout the first decades of the 19th century, the three countries set about consolidating their territorial possessions. An 1825 agreement between the Hudson's Bay Company and the Russian American Company confined Russian activities to the region north of Vancouver Island (54° 40'). A similar arrangement between the Hudson's Bay Company and the American Fur Company proved more difficult to settle. In 1818, representatives of Britain and the United States signed a Convention of Commerce giving citizens of both countries access to the territory west of the Rocky Mountains. The agreement was renewed in 1827, but by the 1840s it had collapsed. In the 1840s, the British were more aggressive in the region that was by this time called the Oregon Territory, and they established a major trading post there. As a result, the HBC controlled much of the coastal trade. British presence did not deter Americans from pressing for access to the region, access they claimed was their legal right based on the terms of the old Convention of Commerce.

American desire for control of the Oregon Territory played a decisive role in the election of James Polk as United States President in 1844. He was swept to victory on a platform that included the re-occupation of Oregon; "Fifty-Four Forty or Fight" was the slogan. But Polk also wanted Texas, and the prospect of fighting both the British and the Mexicans with an army that was under-manned and poorly trained proved daunting enough for the President to modify his ambitious northern plans. When informed that the British were flexible on the placement of the boundary, Polk opted for negotiations.

British flexibility on the issue was born of pragmatism. There was a time when its officials stood fast on an American boundary at the 42nd parallel, the northern border of California, but that was in the early part of the 19th century when the region around the Columbia River was still an untapped reservoir of furs. Since then, the area had been pretty well trapped out. In 1845, the Hudson's Bay Company moved its main post from Fort Vancouver (on the Columbia River) to Fort Victoria (on Vancouver Island). Without pressure from the powerful trading empire to retain control of the region and anxious to maintain a good commercial relationship with the United States, the British government decided that compromise was also in its best interest. The result was the Washington Treaty of 1846 that established a border through the Oregon Territory at the 49th parallel. The line was uncontroversial, likely because it was merely an extension of the international boundary between the Lake of the Woods and the eastern foothills of the Rocky Mountains, as drawn in 1818. The mainland British territory continued to be known as New Caledonia, while the adjacent American possession soon came to be called Washington Territory.

Upon relocating to Fort Victoria, the Hudson's Bay Company entered into negotiations with the British government designed to solidify their trading monopoly. Although they had dominated trade in the region for decades, the company sought a contract that would provide greater legal security for their interests. An agreement was reached in 1849, whereby the HBC was granted proprietary rights to Vancouver Island for the following decade. The grant came with conditions. The British government retained civil authority over the island in the person of a governor whom it would appoint. The HBC also agreed to establish a colony comprised of British citizens within five years. Failure to do so would result in the revocation of the grant.

Hudson's Bay Company officials were well aware that the settlement requirement was one they would not comfortably be able to meet. The simple fact was that the fur trade and settlers did not mix. Settlers needed to have the land divided up and ploughed for farming, whereas fur traders required that it be left in its natural state. James Douglas was also aware of other problems. As early as the 1830s, he had written to inform his superiors that a colony could only flourish with equal laws, free trade and "the accession of respectable citizens," all of which were a detriment to the fur industry. George Simpson, Governor of the HBC in British North America, adopted a cynical and business-like approach on the matter. He observed that even if the HBC did not meet its obligations, it would be of little matter as the region was likely to be trapped out of furs and of no value to the HBC within a decade. In fact, that day already had been anticipated by the HBC. Increasingly its profits came from the new business of supplying the mining communities in California.

The Hudson's Bay Company did advertise in Britain for settlers, but the expense of relocation and the ignorance that clouded peoples' understanding of the far-flung outpost of the empire resulted in little more than a handful of immigrants in the mid-1850s. Nevertheless, the community grew. The HBC was not adverse to its employees (both active and retired) settling outside the fort because they were a known quantity seen as unlikely to disrupt HBC activities. Soon HBC men followed Douglas' lead by acquiring adjacent lands. Douglas considered the land to be owned by the local Native population and he signed treaties with them before allowing its purchase. His actions were a far-sighted approach that minimized conflict with the Natives.

While the settlers included a broad range of social and economic classes, from servants to the officer class, they shared a common cultural heritage almost to a person, a fact that was foundational to the shaping of early Victoria. Since the community was relatively isolated and therefore not readily subject to external influences, it came to take

George Simpson (1787–1860), "the Little Emperor," was governor of the Northern Department of the Hudson's Bay Company from 1821–1826. Simpson supervised a dramatic reorganization of the Hudson's Bay Company's infrastructure after its 1821 merger with the competing North West Company. He trimmed the company's operating costs by cutting staff, consolidating trading posts and eliminating unnecessary expenses. And he did not make all these changes from behind his desk. Simpson went on numerous excursions while supervising the company's outposts, tirelessly traveling by canoe and horseback across the grand expanse of his domain. His efforts did not go unnoticed. In 1826 he was appointed governor of the HBC's trading territories in British North America, and in 1841 Queen Victoria knighted him.

on the characteristics of a town one might find in the English countryside. In fact, it was a contemporary cliché to suggest that Victoria was the most typically British town outside of the motherland. As schools, churches and local government emerged, parallels with Victorian society were reinforced. The appearance of such institutions also ensured that the transition from frontier to civilization was orderly, at least until gold was discovered.

James Douglas first informed his government superiors of the discovery of gold deposits on the mainland in April 1856. The findings multiplied over the succeeding months. It's likely, however, that Douglas was aware of the presence of such deposits for many years before since local Natives, particularly the Couteau and the Salish, had occasionally paid for trade goods with gold. But the actual amount of gold could not be determined. All that could be said with any certainty was that there was gold in the Upper Columbia River, Fraser River and Thompson River districts. The mouth of the Columbia River straddled Washington and Oregon Territories (and formed the border between the two for some 250 miles), but eventually it veered dramatically north, linking with tributaries in New Caledonia. The Fraser River opened into the Strait of Georgia directly east of Vancouver Island in British territory, and the Thompson River joined with the Fraser about 150 miles inland.

As late as mid-1857 Douglas still couldn't speak with authority on the size of the deposits. Despite the lack of information, he was certain that even the limited knowledge about what gold there was would attract prospectors. He was no expert on gold rushes, but even a neophyte knew that rumors were more than enough to set one in motion. He didn't relish the prospect because a gold rush was sure to present problems to the local administration, in other words, problems that would quickly become his.

In 1851 Douglas had replaced the colony's first Governor Richard Blanshard, who, distressed over the local power of the HBC and disillusioned at the amenities available in the small colony, had quit after less than a year on the job. Holding both positions as Governor and Chief Factor of the Hudson's Bay Company gave Douglas considerable authority in the colony. Although, by 1856, there was an elected legislative assembly, most of its members shared close links with the HBC, so the institution proved little threat to Douglas' command. Furthermore, there was minimal external political interference, since communication with British officials often took months. Recognizing the limitations imposed by distance and the speed of events during a gold rush, authorities left it to Douglas' discretion to determine the best way to preserve order. It proved a sound decision.

Douglas was unwavering in his efforts to ensure that the gold rush would be a well-regulated affair. It wasn't just that order appealed to his British sensibilities. Douglas was a cunning and farsighted administrator, characteristics well illustrated in his gold rush plans. His guiding principle was that the mainland remain British. Should there be disruptions in the region, her sovereignty would be in peril. To minimize disruptions, Douglas would ensure that law and order flowed with the force of the Fraser through the region, despite the fact that, at this point, he actually possessed no governing authority over New Caledonia.

When the first wave of prospectors arrived, New Caledonia was considered by British officials to be a separate territory. Douglas did not look upon it that way. His actions were always designed to address the gold rush as it pertained to all British territory in the Pacific Northwest. Some complained that the policies he developed overstepped his command. The complaints quickly dissipated in the summer of 1858, when the colony of British Columbia was created to replace New Caledonia. Douglas was appointed Governor of both colonies. Unfettered by legalities, he was in a position to tame the whirlwind that was the Fraser Gold Rush.

༺༻

Residents of Victoria knew about the California Gold Rush. For years, they had raised eyebrows and tut-tutted at the stories of travelers that depicted American mining towns and gold fields as violent, immoral and generally chaotic. Of course, most were hardly surprised at such stories, for they had long looked down their smug noses at Uncle Sam, the black sheep of the family. They were comfortably confident that such goings-on were a distant phenomenon that could never be witnessed in British territory. Even as they sat on their porches or assembled to watch the *Commodore* sail into port, their leisurely pipe-smoking was mostly infused with a bemused curiosity rather than concern. The attitude was to change as the vessel's passengers disembarked. Pipes clattered to the ground as jaws dropped; the residents of Victoria were in shock.

By the time the vessel was unloaded, some 450 newcomers milled about the waterfront. The number was more than the population of Victoria! Some of them were British, but among the other nationalities were Germans, French and plenty of Americans, including some Blacks. Not only had the population of Victoria suddenly doubled, the solidly British community had been quickly rendered multicultural. And the scuttle-butt was that these newcomers were the dregs of American society, a shiftless, self-interested lot. It wasn't long before the residents' discontent manifested itself in grumbling directed at the Governor. Why had Douglas allowed their peaceful outpost to be disrupted by the likes of this riffraff?

Had he been inclined to offer a reply, Douglas would have responded that it was necessary to funnel prospectors through Victoria to keep the territory British. If Americans were permitted to enter the mainland uncontested, then it would not be long before there were demands that the United States annex the region. Should the few British fur-trading residents in the territory be dramatically outnumbered, and this seemed inevitable, such demands might well elicit action. After all, the memories of "Fifty-Four Forty or Fight" still echoed.

Douglas didn't need to exert himself in defense of his position on the matter. Within a few weeks, the shock expressed by the locals subsided. Reports about the dubious character of the newcomers were overblown. Sure, there were ex-cons, pickpockets and peddlers of vice, but most were far from being the dregs of California's society. Douglas himself expressed surprise to his superiors. "Their conduct while here would have led me to form a very different conclusion; as our little town, though crowded to excess with this sudden influx of people, and though there was a temporary scarcity of food, and dearth of house accommodation, the police few in number, and the temptations to excess in the way of drink, yet quiet and order prevailed, and there was not a single committal for rioting, drunkenness, or other offences during their stay here." Of course, their stay in Victoria was short-lived. Of the first party to arrive, most left for gold country within days.

It was less the newcomers' character that swayed local public opinion than it was the realization that they could make a profit. A quick glance at the ill-equipped prospectors suggested that the supply business could be lucrative, perhaps even promising fortunes as great as what might be discovered in the interior. Frowns of concern were quickly replaced by enthusiastic smiles and Douglas surely had difficulty containing his own contentment at the speed with which locals adopted his perspective on managing the gold rush. "The merchants and other business classes of Victoria are rejoicing in the advent of so large a body of people in the Colony, and are strongly in favor of making this port a stopping point between San Francisco and the gold mines, converting the latter, as it were, into a feeder and dependency of this Colony. Victoria would then become a depot and center of trade for the gold districts, and the natural consequence would be an immediate increase in the wealth and population of the Colony." Rejoice indeed! And it was more than a matter of coincidence that Douglas' employer, the well-established Hudson's Bay Company, was in the best position to profit from the sale of supplies to miners. Douglas was not only a thoughtful administrator; he was also a shrewd businessman.

Forcing prospectors to travel through Victoria did not prove as difficult as it might appear. Although there was an old Hudson's Bay Company route cut through the Oregon Territory, access through the region in the first months of the rush was out of the question. In late 1847, the Cayuse Natives, who lived to the east of the Cascade Mountains and just to the west of the Columbia River, reacted to the increasing number of immigrants by killing a party of 14 Methodist missionaries. The American government was not in a position to mount an armed response of any size against the Cayuse and so the Whitman massacre, as it was called, went unpunished. Until the Cayuse and their allies were brought under control, the Oregon Territory was essentially closed off. The only other land route was overland from the east across the Rocky Mountains. That route was impractical since the range was mostly *terra incognito* and mountain passes were for the most part unknown. More than that, prospectors who chose the route were required to get their supplies in Fort Garry (the closest European settlement on the plains), thousands of miles to the east. They then had to travel for months through some of the most hostile land in British North America, Blackfoot territory. Some attempted this route during the Cariboo Gold Rush, but it was not a factor in the rush of '58.

There was always the possibility that prospectors might try to by-pass Victoria by landing directly at the mouth of the Fraser River. Douglas had anticipated this possibility and had arranged for two ships of the Royal Navy to patrol the region. H.M.S. *Satellite* and H.M.S. *Plumper* churned the waters of the Strait of Georgia ably assisted by the *Recovery*, a merchant vessel chartered by Douglas and manned with revenue officers and sailors from the *Satellite*. Douglas made a request to British officials for additional assistance, at least until colonial administrators were in a better position to enforce the law on the mainland. His request was denied; Britain didn't expect to make a profit on the colony and so was not inclined to spend money there. Nevertheless, the ships fulfilled their purpose so there were precious few prospectors headed for the Fraser River whose first port of call wasn't Victoria. And there were plenty who came, lured north as much by glittering news reports as actual discoveries of gold.

Journalists had accompanied the first wave of prospectors and they were quick to send back appetizing reports readily devoured by hungry readers. "There is no doubt in the mind of any of us as to the existence of gold and the richness of the deposits. The field, too, is very extensive, and I do not believe its limits are one half known yet," suggested a correspondent for the *Daily Alta California*

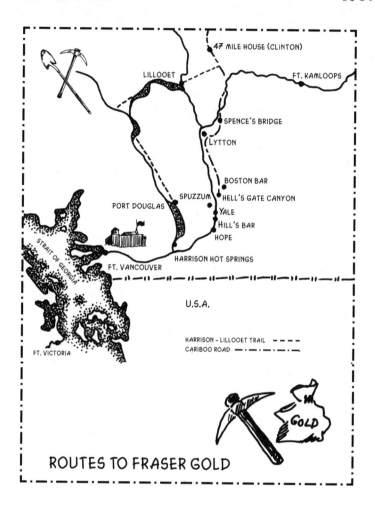
ROUTES TO FRASER GOLD

(San Francisco). His observations apparently mirrored Douglas' information. The rock formations, the veins of quartz, and the accumulations of sand and gravel extending from the base of the mountains to the shores of the Fraser were similar to the geological characteristics found in the gold districts of California. The widely read *Harper's Magazine* printed an interview with a northern-bound prospector. His words were infused with the hopes shared by others taken with gold fever. "Where else in the world could the riverbeds, creeks and canyons be lined with gold? Where else could the honest miner 'pan out' $100 per day every day in the year?"

Such words were infectious and well confirmed by a California correspondent's report that appeared in the *Daily Scotsman* (Edinburgh). "The fever all over the State is intense, and few have escaped its contagion. From Yreka to the north, bordering on Oregon, to San Diego in the extreme south, the masses are in commotion; and from Shasta, one of the northern, to Mariposa, one of the southern mining countries, miners and others are flocking to San Francisco in thousands on their way to New Caledonia." And when, in June, a San Francisco correspondent for *The Times* (London) finally christened the region the new El Dorado, doubting Thomases doubted no more and the full force of the rush was unleashed.

Following the appearance of the *Commodore*, ships arrived with great regularity. By summer they were hardly a matter for local comment, but the rapid increase in population must have elicited some discussion. In 1855 there weren't more than 700 non-Native inhabitants in all the British holdings in the Pacific Northwest and the majority of those were in and around Victoria. By spring, Victoria had swelled by 10,000. In June alone San Francisco bade farewell to 7100 prospectors, who shipped north. As 1858 drew to a close, there were 30,000 prospectors on Vancouver Island and in British Columbia.

The population explosion transformed Victoria with a speed that left residents giddy. Amid the constant rasping of saws and the clattering of hammers, one six-week span in the summer of 1858 witnessed 225 buildings constructed. Nearly 200 of those were stores and a good quarter of those belonged to jobbers, newcomers who had no intention of crossing the Strait of Georgia. Many of those were merchants who had supplied miners down in California and their experience assured that they were soon among the most successful businessmen in Victoria. But there were so many prospectors that even men who hadn't worked behind a counter before did well. When Victoria was declared a free port, massive amounts of goods began to flow in from the United States. So also did big American trading firms, the first of which was started by Peter Lester and Mifflin Gibbs. The Hudson's Bay Company trading monopoly soon dissolved but, with its historical links in the interior, its bottom line hardly suffered.

As Douglas had noted, accommodations came much more slowly. Hotels were expensive and time-consuming to build. As a result, lodgings tended to be of canvas rather than wood. Almost overnight, tents circled Victoria providing temporary shelter for the more fortunate. Many others slept under the stars, in the bushes or on the streets. It wasn't all romance; the loaded revolvers and dirks tucked into each man's belt were a testament to that. If they were lucky, or actually had the skill or guts to use the weapons they packed, their belongings weren't stolen by thieves or lost to the wily overtures of a con-man. And if they had the willpower, they didn't blow their stakes on alcohol, gambling or prostitutes, each of which could be found in large quantities in the boom town.

As often as not, however, prospectors arrived in Victoria without the money necessary to continue on to gold country let alone enjoy Victoria's offerings. In addition to buying equipment, they had to pony up the $5 fee for the annual mining licenses, as well as ensure that they had enough money to buy supplies upriver, and these weren't cheap. In addition, transportation to the mouth of the Fraser was a steep $20. Some turned to negotiation and managed to buy or rent a boat from a local or a canoe from one of the Natives. Others took to building their own craft, giving the Strait of Georgia a hodge-podge of bobbing vessels the likes of which it had never seen. Though the strange sight was visible throughout 1858, it eventually became common to board a ship in San Francisco, stop briefly at Victoria to obtain the necessary licenses and permits and then continue on to the Fraser. Given the cost of living in Victoria, the decision to move on was probably a wise one by those who made it. First class cabin passage by steamer from San Francisco ran $65, while a berth on a sailing vessel could run as little as $25. Whether a miner counted on money, daring or skill, the determined eventually reached the Fraser River, where the adventure really began.

Traveling up the Fraser River *wa*s truly an adventure. The river's namesake, Simon Fraser, had first explored it some 50 years before the gold rush. If prospectors had been aware of his record, likely only the most stubborn would have attempted the trek. Describing what would eventually be known as the Fraser Canyon, a 100-mile stretch between the fork of the Fraser and Thompson Rivers and what would be Fort Yale, Fraser noted, "The water which rolls down this extraordinary passage in tumultuous waves and with great velocity had a frightful appearance; however, it being absolutely impossible to carry the canoes by land, all hands without hesitation embarked upon the mercy of this awful tide. Once engaged, the die was cast, and the great difficulty consisted in keeping the canoes clear of the precipice on one side, and the gulfs formed by the waves on the other, then skimming along as fast as lightning." Fraser continued on to describe stretches where canoeing was impossible and overland travel almost so. "An Indian climbed up to the summit, and by means of a long pole drew us up, one after another. This work took three hours. Then we continued our course up and down, among hills, and along steep declivities of mountains, where hanging rocks and projecting cliffs at the edge of the bank of the river made the passage so small as to render it difficult at times, even for one person to pass. In places we were obliged to hand our guns from one to another, and the greatest precaution was required to pass singly free of encumbrances.... We had to pass where no human being should venture." But miners proved to be a stubborn lot, and it would take more than geography to deter them.

For the first 120 miles, the Fraser is a deceptively tranquil river. But when it reaches the coastal mountains, it becomes terrifying as rushing streams tumble down the mountainside to feed the stretches of white water that roar through deep gorges. In spring and early summer, run-off from melting snow high in the mountains caused the river to rise as much as 100 feet above low level. Getting to the gold fields proved so difficult that it left some miners without adequate words to describe the trip, as is evident enough in the letter Franklin Matthias wrote to a friend. "We have arrived in these mines at last, after one of the hardest trips on record. I shall not attempt to give you a narrative of the difficulties and dangers of traveling on this river, as it would be impossible for me to do justice to it...." Although a stretch named Hell's Gate tends to speak for itself, James Douglas' pen was more up to the task of describing the tribulations miners encountered while trying to travel up the Fraser. "Many accidents have happened in the dangerous rapids of that river; a great number of canoes have been dashed to pieces and their cargoes swept away by the impetuous stream, while of the ill-fated adventurers who accompanied them many have been swept into eternity." Estimates suggested that one in four canoes were lost while trying to make the journey upriver.

There were prospectors who saw goods swept downstream and still didn't think twice about their decision to join the gold rush. But the physical environment wasn't the only obstacle that confronted eager miners. The mainland was peopled with Native tribes, most of whom didn't want the white interlopers anywhere near their land. As was quickly discovered, the more aggressive Natives were willing to kill if necessary, so when prospectors saw scalped or decapitated bodies following the Fraser's powerful current to the Strait of Georgia, some second guessing was clearly in order.

For decades, the mainland Natives had been on good terms with their European trading partners. It was easy enough to get along with the Hudson's Bay Company, because they didn't want the Natives' land, only their furs. Even when HBC employees first became aware of the gold deposits in the mid-1850s and began to extract the metal, the Salish who lived in the area were mostly cooperative. The men of the HBC were a known quantity and few old friendships were shattered over gold. The attitude of the Natives changed with the arrival of Yankee prospectors.

The newcomers weren't content to let the Salish and Couteau mine for gold as they had long done. Likely the miners found it odd that the Natives were even interested in gold, as such had not been their experience in the United States. Nevertheless, competitors or not, the miners had their own way of dealing with Natives and rarely was it with friendship. James Douglas anticipated such violence and was sympathetic to the Native situation when the bloodshed finally came to pass. He wrote as much to his superiors in July 1857.

"A new element of difficulty in exploring gold country has been interposed through the opposition of the native Indian tribes of the Thompson River, who have lately taken the high-handed, though probably not unwise course, of expelling all the parties of gold diggers, composed chiefly of persons from the American territories, who had forced an entrance into their country. They have also openly expressed a determination to resist all attempts at working gold in any of the streams flowing into Thompson's River, both from a desire to monopolize the precious metal for their own benefit, and from a well-founded impression that the shoals of salmon which annually ascend those rivers and furnish the principal food of the inhabitants, will be driven off, and prevented from making their annual migrations from the sea."

Gold and salmon aside, Natives had other reasons to resist the encroachment of American prospectors, the most pressing of which was self-preservation. In the early days of the rush, it was common to hear northern-bound Americans declare that they would rid the gold country of Natives. Their history suggested it wasn't an idle threat. Indeed, the boasts were often acted upon. An unarmed Native had a good chance of being shot in the back for his troubles and, while large scale attacks were rare, even numbers were no guarantee of safety in the face of angry miners.

Once gold was discovered along the Fraser and beyond, it was just a matter of time until the waves of prospectors rolled inland and put an end to the way of life enjoyed by the Natives of the region. For many years prior to the rushes, Natives had used gold in payment for supplies at the Hudson's Bay Company posts. Little was made of this information until the *Otter*, carrying 800 ounces of the yellow ore, made its way to San Francisco in 1858. It set off a spark that ignited the entire west coast in a frenzy of gold rush fever. Although the Natives were instrumental in first finding gold in the Rockies, none profited by it. Many left their nomadic existence to work for pittances, the men on the sluices and in the mines, and the women as prostitutes. The Natives in this picture are a group of Salish. They lived in the vicinity of the Fraser River.

A particularly vile incident occurred on the shores of Okanagan Lake, in British Columbia. A party of miners had followed an old Hudson's Bay Company trail up through Washington Territory. Apparently, Natives on the American side of the border had harassed them, so when they found provisions at an unattended Native village on the lakeshore, they destroyed the goods. The following day, when the Natives returned, the miners massacred them. One of the mining party reported that the Natives were without weapons and further admitted that the whole affair was brutal.

For a time, Natives were content to disrupt mining activities by stealing whatever tools, goods and stock they could get their hands on. Such was tit for tat, since the miners were, after all, stealing their gold. But as the violence perpetrated by the miners became bloodier, many Natives responded in kind. One of the first episodes occurred near the Fraser Canyon in early August 1858, when a party of Natives ambushed a 400-packhorse caravan. They left three men dead and stole 100 horses. The Natives followed up their victory with a guerilla-type strategy, sniping miners from a distance and retreating into the forest. It wasn't long before the short-fused Americans had had enough and they formed a vigilante committee to track down the murderers. Americans and Natives met at Spuzzum, about 10 miles north of Yale. Seven Natives were killed and the remainder of their party scattered in the face of the ongoing attack. The Americans then turned their attention to the dead. They had discovered the Natives' burial ground and burned the intricately carved tombs to the ground. As they watched the flames dance, the vigilantes were joined by a second party, making some 200 in all. They traveled up the Fraser, crazily focused on finding more Natives and making good on their threat to clear the gold country of their presence.

The vigilantes stumbled on a party of five injured miners, who informed them that Natives had attacked them. The rest of their mining party had been less fortunate. More determined than ever, the horde continued on after the Natives. In late August they found them. The rumors that floated down the Fraser River gave every indication that they surely wished they hadn't. Forty-two white men were murdered; the bodies of 38 were terribly mutilated. Many in the white community were incensed at the news and, for a time, Natives had an especially good reason to fear for their lives. Throughout the lower Fraser, Natives were assaulted, killed or reduced to powerless bystanders as their villages were destroyed.

The news of the deteriorating situation eventually reached James Douglas, who decided that the best course of action was to personally address the matter. In early September, he headed for Yale accompanied by a 35-man military contingent. En route he received further intelligence suggesting that the reports of white deaths were greatly exaggerated; there were, in fact, only two lives lost. Nevertheless, the violence between Natives and miners had reached such a pitch that Douglas decided it was best to continue on and to resolve the situation for good. By the time he arrived at Yale, the spate of violence had played itself out.

Douglas carried out his investigation and it was soon clear to him that the cause of the troubles were the miners, the most "ruffianly looking men" he had ever set eyes upon. Their appearance was not deceptive. When the Governor addressed the alleged culprits, insisting that they obey the rule of law, his advice was met with rude responses and harsh comments questioning British authority in the territory. One of the peculiarities of the Fraser Gold Rush was that Americans dominated it. While there were those from other countries, many who might have come preferred to try their luck in either Australia's gold rush or Nevada's Comstock Lode silver rush. Douglas had long feared that anti-British sentiment might boil over and he was steadfastly determined to place a tight lid on it. He sent for the Crown Solicitor of Vancouver Island to preside over a murder trial. Those accused and subsequently found guilty were banished from the region since there was no jail to accommodate them.

As 1858 drew to a close, Natives and miners reached agreements that kept the peace. It was a matter of necessity. The Natives weren't readily going to give up their land (although they watched their traditional territories shrink drastically during the gold rush) and the miners were not going to retreat. The result was an uneasy co-existence. There continued to be the occasional violent outbreak, but they were rare during the Fraser Gold Rush. While some Natives continued to work their own claims, increasingly they took to working for the newcomers. Some were employed in the larger mining operations, but more commonly they were employed as farm labor, packers or transporters. It was nearly impossible to find whites who would take on such positions since the gold fields provided competition that was far too attractive. But it was mostly seasonal work, and all too often the Natives weren't paid. By 1860 poverty characterized the plight of most mainland Natives.

To improve their situation some Natives turned to the growing towns, where they could more easily find work or sell their game, fish and crafts. Generally they didn't live in the settlements, but even an adjacent Native presence was too close for most whites. Residents loudly protested; Natives were immoral and their vices eroded the good behavior of the local population. Certainly there were problems, though often as not, their roots were to be found in the white community. Alcohol was readily available and booze peddlers made a handsome profit selling it to the Natives. Crime increased, likely as a result of the diminished restraints due to drinking. Often the more violent outbursts were among Natives, since the towns attracted different tribes, many of whom were traditional enemies.

Native women were also involved in prostitution, which flourished. Wealth was an important indicator of status among many west coast Natives tribes and when they discovered that there was easy money to be made in the profession, it was easy enough to gravitate towards it. The most lucrative locale for the activity was Victoria, where a soiled dove could earn twice as much money as she might up-river. By the early 1860s, "squaw dance houses" featured Native women dressed in the paints, powders and finery commonly seen in prominent California brothels.

As the decade came to a close, what gory violence there was of the early days of the Fraser Gold Rush was mostly a memory. Given the antagonistic behavior and racist attitude of many prospectors, the fact that the gold fields weren't red with the blood of the slaughtered was due in great part to the foresight of James Douglas. His insistence on law and order did much to ensure that mining was actually the main activity on the mainland.

∽∾

The rain came down hard and the dark sky and cool wind made it seem as if it was late fall and not midsummer. A handful of tents was pitched alongside Summit Lake, about halfway between 8 Mile House and Port Anderson on the Harrison-Lillooet Trail. In one of the tents sat a man reading a book. A pipe protruded from between a moustache waxed to sharp points and a full Vandyke beard. The powerful odor of the tobacco had the pleasant effect of overpowering the smell of dank canvas. The man slid his finger up to the corner of the page and tried to turn it, but the paper was damp and he had to fumble with it. Traveling through the treacherous gold country of the Fraser River was a tough assignment and try as he might, he couldn't stop his books from becoming mildewed! Judge Matthew Baillie Begbie sighed and tossed the book aside; Plato would wait for another day.

Begbie stood, his six-foot plus frame dominating the interior of the tent. He went to its entrance and peered out. As quickly as the rain had started, it stopped. He called to his Native guide to prepare his tackle. He'd do some fishing.

There were many odd characters who followed the gold rush west to British Columbia in the late 1850s, but there weren't many as unusual as Matthew Baillie Begbie. There were even fewer who'd have the impact he did on the region's development. With a degree from Cambridge and years of experience as a Chancery lawyer in his pocket, Begbie arrived in Victoria in November 1858. He came to fill the post of Judge of British Columbia and he wasted little time putting his mark on the nascent colony, departing for the mainland colony within weeks.

The position of Judge of British Columbia was the lynch-pin of James Douglas' strategy for ensuring the mining territory was a lawful place. From the first, Douglas envisioned that the law would have a visible presence in gold country, because he considered that the most effective way to address the inevitable mining violence. Since British Columbia was without an elected or appointed body that could share in policy-making, he created through decree a multi-layered legal system. At the top would be a judge with supreme control, who would ride the mining region as an assize circuit, dispensing justice as he saw necessary. His efforts would be supplemented by local Justices of the Peace (who were charged with hearing only lesser crimes and minor litigation, and whose decisions were subject to Begbie's approval), Gold Commissioners (local authorities who registered mining claims, collected revenues, settled disputes and generally developed the regulations that governed mining communities and who often held the joint appointment of justices of the peace) and a small corp of appointed constables. During the Fraser Gold Rush, these peacekeepers and lawmen usually didn't exceed 40 or 50, yet they were responsible for thousands of prospectors mining in a region of many hundred square miles. It proved adequate and few honest men feared for either their lives or property.

Establishing the position of Judge of British Columbia and finding the proper man to fill it were two different matters. The man for the job had to accept the loneliness of a truly challenging frontier life. He had to trade in the fancy carriages and estates of English society for leather saddles and disagreeable hotel rooms. He had to be prepared to face hostility from every quarter, including miners, Natives, businessmen and assorted lowlife swept along in the gold rush. He would have to be dedicated and resourceful, with an understanding of the law that was subject to self-doubt in neither knowledge nor application. The candidate would be helping to establish a new society, so just anyone would not do.

Matthew Baillie Begbie was not just anyone. At 39, he was already fluent in a handful of European languages both contemporary and ancient and he wasn't long in British Columbia before he mastered several Native dialects. Whether he was holding court under a tree or in a more

Appointed as Judge of British Columbia in 1858, Matthew Baillie Begbie was faced with a daunting responsibility. He was charged with carrying out justice in the remote gold fields of British Columbia, where a large population of desperate fortune seekers threatened to turn the British colony into a lawless American territory. Begbie traveled long distances throughout the region, holding court in tents, bars, under trees, ensuring everyone was aware that they were on British soil and, as such, were subject to the crown's laws. Standing six and a half feet tall with a stern countenance and a burly frame, the looming judge made quite an impression on the men who worked British Columbia's inland waters. But the "Hanging Judge" was both fair and scrupulous, and he earned the respect of the rough men in the Northwest. He eventually settled in Victoria, where he spent the remainder of his days steeped in the trappings of English gentility, his faith in British civilization unaffected by his rough tenure on the western frontier.

proper shelter, he used his facility with languages in a most thunderous and unsettling manner to both the accused and other onlookers. Regardless of the setting of the place of judgment, Begbie always donned his black judicial robes, thereby adding to the show. He quickly became known as a hanging judge, though he did not sentence anyone to death whom the jury didn't convict. In his defense, the only punishment for murder was death. His reputation was such that many an alleged criminal fled rather than suffer the punishments Begbie was known to mete out.

Begbie knew the law well and was inclined to forcefully impose it, as demonstrated early on in his first circuit. At Fort Yale he was presented with two Natives charged with murdering an American. Upon discovering that their detainment was based on suspicions alone, Begbie had them released. His justice would be based on hard evidence; skin color and stereotypes would have no place in his court. It was a philosophy that was also apparent in his treatment of the Chinese, many of whom eventually participated in the rush. To the displeasure of some, it took more than discrimination for Begbie to convict. If people didn't approve of his decisions, then he had unorthodox strategies to deal with them, as the companions of a man he had convicted discovered in a most disagreeable manner. While in his hotel room relaxing after a court session, Begbie overheard the two plotting revenge against him. Only shooting was good enough for the man who brought down their pal. Armed with a loaded chamber pot, Begbie exploded into their room and dumped its contents over them!

Despite the power that accompanied his position, Begbie would not overstep his authority. If he presided over a jury trial, he stood by the jury's decision even if he disagreed with it. He would not, however, mask his displeasure. On one occasion, a Victoria jury released a man who had allegedly killed a miner. Begbie was convinced of the man's guilt, but set him free with a few caustic words of wishful thinking. "Prisoner at the bar, the jury have said you are not guilty. You can go, and I devoutly hope the next man you sandbag will be one of the jury."

Douglas came to depend on Begbie, and not only for his rigorous assertion of the law. Begbie spent long hours making detailed notes on the terrain of the mainland interior, suggesting locations for roads, bridges and town sites. Such information was invaluable to Douglas because part of his strategy for controlling the gold country was to make it more accessible, but only from the appropriate point of entry. By 1859 the United States army had effectively ended all Native resistance in Washington Territory, and local businessmen began promoting a line to the Fraser gold fields from Whatcom (Bellingham), just south of the border. The Whatcom Trail linked up with an old Hudson's Bay Company route. Enthusiastic prospectors turned sour on the trail when it became known that it took two months to travel. Nevertheless, it provided Douglas with another reason to act to ensure that the mainland remained British. He'd set up a transportation route that ran west-east and make it so efficient that no miner would want to make the effort to come from the south by land.

The British government was receptive to the idea so assigned the Royal Engineers the task of surveying and upgrading routes. The Columbia Detachment of the Royal Engineers (sappers) were an elite branch of the military, specially chosen for the task because of their "superior discipline and intelligence," characteristics that made them less likely to abandon their assigned duties for a chance at prospecting. (Others in the military were paid higher wages to keep them from deserting; the sailors aboard the *Satellite* and the *Plumper*, for example, were paid $1 a day, a considerable wage for the time.) The engineers were directed to downplay their military credentials so as not to irritate the many American prospectors. Nevertheless, Douglas clearly saw their armed presence as a valuable show of British force in the mining camps. Commanded by Colonel Richard C. Moody, the first of the contingent (those fortunate enough to take the Panama route) arrived in November 1858, after a two-month journey. The bulk of the 180-person force (there were about 20 women) traveled around the Horn and the last of them didn't arrive until the following spring.

The sappers were responsible for upgrading old roads, building new ones, constructing bridges and surveying town sites. Moody had left Britain with orders to keep costs down, but he wasn't quite successful because engineering in the mountains was an expensive undertaking. There were also questions about Moody's own scruples, rumors that he was unnecessarily increasing costs and lining his own pockets. He profited from the efforts of the Columbia Detachment by purchasing nearly 4000 acres of prime land at less than $2 an acre. Evidently superior discipline and intelligence provided no sure tonic against the many strains of gold rush fever. His apparent greed was only one of the reasons why he had a quick falling out with Douglas. The other was his opinion of Begbie's surveying efforts and drawings, which he characterized as amateurish.

The Royal Engineers were to have their greatest impact on the Cariboo Gold Rush of 1863, when they successfully conquered the hostile terrain of steep, jagged mountains and deep-valleyed rivers to give easier access to the inland diggings. But they cut their teeth closer to the coast, first putting their efforts towards linking New Westminster (which had been selected as the capital of the mainland colony of British Columbia), at the mouth of the Fraser River, with the Burrard Inlet. At Douglas' insistence, they then turned to what was probably their most important work during the Fraser Gold Rush, the Harrison-Lillooet trail. The route connected southern (Port Douglas on Harrison Lake) and northern (Lillooet, three miles north of Seton Lake) points on the Fraser River, while bypassing the worst of its rapids. The sappers didn't start this job from scratch because the route was first carved in 1858, an accomplishment in itself, with a story to match.

In the early months of the rush, most prospectors saw their dreams become nightmares at Fort Yale, as the glitter of gold was obscured by torrential white water. The upriver stretch of the Fraser through Hell's Gate and the Canyon was simply too challenging and too dangerous for many to tackle. It was a tough pill to swallow because the prospectors knew that the small amounts of gold at and around Hill's Bar and Yale had to have been washed downstream from larger deposits. So the prospectors, faced with the daunting obstacle, etched a route around it. By late July 1858, 500 (divided into 20 companies of 25 men each) were busy working on the route. In fact, not only were they volunteers, they were required to pay up front to participate in the work. Each man gave $25 as security for his good conduct. When the work was done, they were repaid in kind and were given $25 worth of supplies at Victoria prices. Ultimately, they made a little profit on the job, but clearly that was not their motivation. The prospectors were of a unique breed and would do whatever was necessary to keep their dreams alive. Starting the task with little more than a few old Native trails, the road, 108 miles long, was completed in October.

The Royal Engineers widened the trail to a 10-foot wagon road. While it might have been a safer route than tackling the Fraser River, it remained an expensive one. Between Port Douglas and Lillooet, freight had to be transferred from steamer to wagon eight times and each time the cost of transportation increased. As a result, by 1861, the sappers had set their minds on conquering the rocky barriers of the Fraser Canyon, a story better told as part of the Cariboo rush.

༄

The inhospitable Fraser River and the determined Natives who lived around it might have been among the greatest obstacles to finding gold in the rush of '58, but they weren't the only ones. A prospector needed money because his was an expensive undertaking. The farther up the Fraser he staked a claim, the more he had to pay for supplies. He had to be careful where he picked up his goods too because life along the Fraser River, despite Douglas' efforts and sometimes because of them, was not always safe. Racial tensions were also thick, and it wasn't just anti-Native sentiment. The many Chinese who worked on the Fraser were often subject to discrimination. More than the water of the Fraser boiled in the late 1850s.

As was the case with most rushes, there were plenty of prospectors on the Fraser who didn't find much gold. The river itself made the mining difficult. There were few sand or gravel bars (the river was usually too high) and it was difficult to dig ditches. Usually men prospected with the rocker, which wasn't the most effective of tools, or they idled waiting for the water level to fall. Reports of miners taking $8 to $10 a day were common; less usual were the stories of men taking $50 to $75 a day. If a fellow was making that kind of money in Toronto or New York, he would have felt pretty good about it. But not on the Fraser, where gold disappeared almost as quickly as it was found.

A miner's claim had to produce $5 a day for him to break even. A letter written by prospector Joseph Haller and sent to his relatives in Pittsburgh illustrates why. "Food is very dear, one pound of flour $1.50, pork $1.75 per pound, beef $1.00 a pound, sugar $1.50, beans $1.45, tobacco $6.00. Everything is dear...." In Victoria, pork was $.25 a pound, beans only $.50 and customers didn't have to deal with unscrupulous traders who soaked their tobacco in water so that it weighed more at the time of sale. But Haller couldn't have been doing too badly because he told his family he had $100 to send them. The only problem was that it cost $50 to mail it and there was no guarantee of its arriving!

The prices reflected the challenges of transportation. While the Harrison-Lillooet wagon road made the region accessible to the mule teams, it remained a pricey route. Though many shipping agencies chose to freight goods on the Fraser, it wasn't because it provided an easier passage. John George Brown, who would become well known in southern Alberta as Kootenai Brown, spent a spell employed by one of the Fraser River freighting companies. He was paid a good salary of $6 a day, but it was demanding work chosen only by those down on their luck. The gear was packed on large cedar canoes—Brown's vessel was 30 feet long, with a 7-foot beam—because the birchbark canoes mostly used by eastern Natives would have been too easily splintered by the rapids. But as often as not, the gear was packed on the backs of the men, as recounted by Brown. "Sometimes the boat would stick and we would have to take the cargo ashore and "pack" it with head straps a couple of miles or more over the sides of mountains impassable to horses and wagons. On these portages we would carry from 150 to 300 pounds according to the strength and endurance of the man." As tiring as that was, it was preferable to challenging the rapids. Brown noted that through the Fraser Canyon the river runs at about 12 miles an hour (by comparison, the Missouri snakes along at a leisurely 2 miles an hour). Brown's stretch was 25 miles, and while it took four and a half days to go upriver, the return journey took two and a half hours! Likely not all chose to return by boat.

The prospectors who kept the freighters in business were a mixed lot. There were plenty of hard-working miners anxious to make the big strike. Despite Douglas' efforts, with the news of gold, crime was as inevitable as the prospectors themselves. There were plenty swept along in the rush who never intended on trying to make an honest buck. In early June 1858, a newspaper correspondent in San Francisco reported that the more-or-less law-abiding men who arrived on the first of the northern bound ships were soon to be replaced by a less desirable sort. "I was informed by a gentleman from the interior [of California], who observed the exodus now taking place with much interest, that the greater part of it was composed of idlers and vagabonds who had been hanging about the different mining towns and camps 'out of luck,' doing nothing, and ready for anything that might 'turn up'...." Of course, what "turned up" was the Fraser Gold Rush.

Douglas' system of law enforcement was mostly in place by the summer of 1858. It was the constabulary that would be the first contact with idlers, vagabonds and anyone else intent on breaking the law, but his first attempt at creating a constabulary was a failure. Douglas formed a militia regiment and chose to fill its ranks with escaped American slaves and free Blacks who came north with the rush. He didn't have to pay them as much as he might white constables and the fact that Douglas' mother was a free Black from British Guyana probably also figured into his decision.

The problem with Douglas' Colored Regiment, as it was called, soon became apparent, that is, white American miners had no respect for the Blacks' authority. Subsequently a more decentralized strategy was adopted in the mainland colony (Vancouver Island had its own police force), with local magistrates appointing their own constables. But even the constables weren't safe from the occasionally crooked arm of the law.

The Fraser River was no California but it had its share of crime. There were robberies, assaults and occasionally murders, as often as not perpetrated by a disgruntled partner as a stranger or a Native. One of the worst flare-ups of the Fraser Gold Rush occurred in late 1858, and although there were local circumstances that set it off, the real cause of the problem was a couple of Douglas' choices as inland magistrates.

As the river flowed, Hill's Bar and Fort Yale were the first two locations where there was a significant amount of gold extracted. Realizing that the population in the area would dramatically increase, Douglas appointed two local magistrates, George Perrier and Captain P.B. Whannell. He must have made the appointments without proper background checks and that was to prove costly. Whannell had misrepresented himself by claiming he was an officer in the Royal Victoria Yeomanry Corps. Allegations of misconduct soon suggested he had few of an officer's attributes. He was a public drunk and skimmed money into his own pockets from the fees paid by miners. He didn't, however, shy away from controversial decisions.

In December, Bernard Rice, a known gambler and all around mean sunnuva-bitch, shot William Foster, a Fort Yale miner. Rice escaped and was kept hidden by friends in Hill's Bar. Whannell sent three special constables down river in an effort to find and capture the suspect. He couldn't find him, but arrested Foster's partner and his servant (who was to be kept as a witness). Hill's Bar Justice of the Peace Perrier was incensed; Whannell had overstepped his jurisdiction. He sent his constable, Hickson, to Fort Yale to let Whannell know of his displeasure. Whannell didn't like either the tone or task of Hickson and had him thrown into jail for contempt of court. Perrier took the action as a slight to his own authority and swore in a group of new constables to break Hickson free and to arrest Whannell.

Hill's Bar had plenty of tough characters. A good few of the vigilante committee that had set off after the Natives a few months before were from the locale. Inevitably, many of the 20 new constables—labeled by some a posse, by others a lawless band of ruffians—were of questionable character. None, however, had the reputation of the group's unofficial leader Ned McGowan. Though he had been a lawyer, politician, newspaper publisher and police captain over the years, McGowan found himself drawn to society's underbelly. Before he left San Francisco, he was called "chief of the vultures," a reference to his leadership of the law and order Vigilance Committee, which was said by some to be the most notorious criminal group in the city.

When Whannell received word of his impending arrest and of the man who was assigned to carry it out, he wrote a hurried letter to Douglas. Perrier and McGowan (whose reputation as a troublemaker had reached Victoria and was well known by Douglas) were characterized in most unflattering terms, but it was the ominous tone of the missive that spurred the Governor to quick action. "This town and district are in a state bordering on anarchy; my own and the lives of the citizens are in imminent peril. I beg your Excellency will afford us prompt aid.... An effective blow must at once be struck on the operations of these outlaws, else I tremble for the welfare of this Colony."

Douglas, who was always on the lookout for a spark that might set off a rebellion, took the threat seriously.

The Governor dispatched the Royal Engineers under Moody, one of only two occasions during the Fraser and Cariboo rushes that they were needed to fulfill a purely military assignment. The mission was expensive and dangerous; travel up the icebound Fraser in winter came with many risks. Before they arrived, Whannell had been arrested and brought before Perrier where he was fined $25 for contempt of court. The subsequent investigation by the Moody determined that the whole situation had been blown out of proportion and that, more especially, there was no reason to fear that McGowan was about to lead a resistance against British authority. Nevertheless, as a result of McGowan's War, as it became known, the two justices of the peace were replaced.

Even honest law officials were hard pressed to combat the racism that came with the gold rush. It found expression against many ethnic groups, as cases involving Natives and Blacks have already illustrated. But none probably suffered harsher discrimination than the Chinese. Although there were some Chinese who were swept along from San Francisco in the early days of the rush, many more (estimated at about 4000) came from Hong Kong around 1860, at a time when most of the two- or three-man mining operations had been abandoned for claims upriver or bought out by larger operations. Many among this second wave of Chinese weren't so much prospectors as they were paid labor. They did much of the hard, dirty work that others wouldn't. They did it quietly and they accepted low wages (maybe half of what white workers expected) without complaining. Employers loved them but plenty more hated them.

Few Chinese were able to speak English and those who could usually spoke with a thick accent that racial sensibilities

Chinese prospectors in the western gold fields had to contend with more than the physical hardships of gold mining. Victims of an acerbic and almost constant racism, the Chinese were made acutely aware that they were not welcome along the gold-bearing streams of North America. The term "a Chinaman's chance" came from white miners mockingly assessing the probability that a Chinese miner would strike it rich. Nevertheless, many Chinese prospectors were successful at finding gold, acquiring a reputation of making lucrative finds on sites that white prospectors believed had been mined out. Many of the Chinese that arrived in British Columbia were prospectors who had moved up the coast from California, still on the hunt for the big strike. Gold-separating machines like the sluice box pictured above made panning much easier.

of the time allowed to be mocked. They were treated as second-class citizens, a position many of their contemporaries saw as appropriate given their practices. Those Chinese who didn't work as employees were usually found cleaning up the bars long since abandoned. Observers noted that they lived on rice and tea and not much else and suggested such habits were anything but normal. As likely as not, white miners were upset because the patient Chinese often discovered plenty of gold while cleaning up and they didn't have to spend a whole lot on supplies. Although government officials, including Douglas, praised their civic-minded efforts in helping to build the Harrison-Lillooet trail, there is little in the way of records to suggest that such an activity did much to warm the hearts of their fellow miners.

The arrival of the Chinese miners signaled the end of the Fraser Gold Rush. Those who could afford to build wing dams, sluices and flumes, and dig extensive ditches were replacing the small operations. Such operations were more mechanical and labor intensive and drew on a new type of miner, one who worked for a wage, rather than for the thrill of a find. By 1860 those hardy prospectors who had decided not to abandon British Columbia finally penetrated the Cariboo region and thereby set things in motion for a whole new gold rush.

At its height, through the late summer and fall of 1858, the Fraser River truly boomed. It attracted some 11,000 prospectors, most working pretty much shoulder to shoulder along a 200-mile stretch of the river. Throughout the harsh winter of 1858 to 1859, only about 3000 of them remained in gold country. Some found solace in winter camps or in the bigger communities of Fort Yale and Fort Hope, but many others found themselves alone or in the company of one or two others on their claims. Many were unprepared; the mildness of the previous seasons had led them to believe winter would be California-like. It wasn't, and the cold months surely saw their share of suffering. The following years miners were less inclined to winter on their claims and it would be many decades before the numbers along the Fraser would reach the heady days of the summer of '58.

In addition to those who actually persevered in gold country, thousands more came. Many of those left when faced with the forbidding challenge of the Fraser, while others in the rush remained in Victoria hoping to make their fortunes there. And fortunes were made. James Douglas estimated that there was in excess of 100,000 ounces of gold extracted from the region in 1858. With the subsequent technical advances, the amount of gold produced in 1859 was some three times that but there were many fewer miners employed. The gold filtered back through the mining towns and eventually to Victoria, filling the pockets of those who never shook a rocker or swung a pick. And, of course, there were fortunes lost, as put poetically by Watson Hodge, an American miner turned restaurateur and storekeeper. His business was at 4 Mile House (4 miles above Fort Yale), but visitors to his place in the summer of 1860 found the door locked and a message tacked to the door, "My whisky's gone, and credit too, and I've put out for the Cariboo. So if you want rum or rye or ale, you'll have to get it down at Yale. (And pay for it.)" Prospectors were nothing if not optimistic; dreams only waited for another rush.

While the Fraser River itself ensured that the northern rush of '58 was much different than its California counterpart, the presence and efforts of James Douglas was no less a differentiating factor. As fierce as the Fraser itself and as obstinate as the rock faces that guarded the flowing waters, Douglas planted a stamp on the region and the activities that filled it in a way that only hindsight would have suggested was possible. At a time when the issue of sovereignty was in flux, in a place so treacherous as to defy the presence of all but the most stubborn, with a people whose inclinations were shaped by unbridled self-interest, Douglas triumphed and ensured the continuing existence of a unique place in the Pacific Northwest.

3
The Cariboo
(1862)

PETER DUNLEVY CROUCHED over his pan, ankle deep in water. He was at the mouth of the Chilcotin River, which forks off from the Fraser River in a westerly direction south of Williams Lake. It was 1860 and he had long since abandoned the older diggings along the Fraser River in his search for gold. He and his four companions had been testing the gravel in the Chilcotin for many days with precious little to show for their efforts. They had turned up some color but nothing much bigger than the head of a pin.

Dunlevy dipped his pan into the water yet again when he felt the hairs on the back of his neck begin to prickle, the kind of feeling a fellow gets when he's being watched. In a splash he leaped up and spun around. There, not more than a few yards away from him, stood a Shuswap.

"Boys!" called Dunlevy. "We've got company."

The men stopped what they were doing and looked to the visitor.

"How long's he been here?"

"Danged if I know. Never heard him arrive."

"S'pose he understands English?"

"Lots around here do. Ask him."

"What you want, fella?"

"Me Tomah. Me with Company. Me watch," replied the Shuswap.

"He's got friends with him!" barked one of the miners. "Better get your rifles, boys."

"Just hold on a minute," said Dunlevy as the men scrambled for their weapons. "If he was out to get us, we'd be dead by now. I think he means he's with the Company, you know, the Horny Boys Club," he suggested, giving the popular slang term for the Hudson's Bay Company. Tomah confirmed his suspicions.

"Ahh, cripes boys, we got better things to do than to pay him any mind."

Dunlevy turned back to his pan, and slowly his friends joined him. An hour later the Shuswap was still there, observing quietly.

"That's the strangest thing I ever seen. Gives me the creeps."

"You think he's waiting to see if we find anything?"

"I gotta think that he doesn't need us to tell him where the gold is. The Indians know; they've been mining it for years."

"By God, you're right. He probably does know where the gold is! Let's ask him," suggested Dunlevy.

Dunlevy motioned to the Native, who responded by walking towards him. When the Shuswap stood close, Dunlevy reached into his pouch and pulled out a small nugget about the size of a grain of wheat.

"Tomah, you know what this is?"

The Shuswap grunted.

"Me see plenty. Bigger dan dat."

"No kidding! Where!"

The Shuswap picked up a twig and scratched a map in the sand.

"Here," he said, marking the spot.

Dunlevy's eyes were as big as his prospecting pan and his heart was racing.

"It's right there, boys! Right there!" Dunlevy's friend shared his excitement, smiling and rubbing their hands together.

"Wait a minute," continued Dunlevy, a frown appearing on his face. "Where the hell is that? Boys you got any idea?"

The points on the map were unfamiliar and none of them could identify the location of the spot. They certainly had no idea as to how to get there. Dunlevy told the Shuswap of their confusion.

"You come to lake just east of here, less than a day's walk. My cousin show you where gold is."

"He's got to mean Lac la Hache, just over yonder," said a rejuvenated Dunlevy, while pointing in that direction. "We'll be there chief. Just got to get some supplies."

The Shuswap stepped away from the map and continued on his way. Dunlevy and his friends went back to Lillooet to purchase what they would need for the trip. Days later they arrived at Lac la Hache, and found a large Native encampment of Carriers, Shuswap and Chilcotin. Their hot-blooded enthusiasm froze into cold-blooded fear. Was it a trap? Had Tomah set them up? Memories of hairless miners floating down the Fraser River during the early days of the rush of '58 reappeared in vivid color. They were about ready to skedaddle when Tomah appeared bearing a great smile. He explained that the gathering was a yearly event, one of games and celebrations. Relief washed over the men and their enthusiasm quickly returned when he introduced his cousin Long Baptiste, who had agreed to lead them to the gold deposits.

Without wasting time, they set off. Baptiste directed them to the Wildwater River (Horsefly River), south of Quesnel Lake and well east of most prospecting. The men broke into large smiles when they saw the gold-bearing character of the banks. Hurriedly ripping their pans from their packs, they ran to the shoreline. It was the spring of the year and the river was running high, hardly ideal conditions for prospecting. But the men weren't deterred. Dipping their pans into the water, they immediately pulled out more color than any of them had yet seen in the region.

The Dunlevy party was the first to discover gold in the Cariboo; they'd be joined by the second within a day. There was no such thing as a secret in gold country and, when Dunlevy and his party left Lillooet with their supplies, they were followed by suspicious miners who were always on the lookout for signs of a strike. Although Dunlevy eventually sold his share in what would be the productive Horsefly Mine and bought a hotel at Soda Creek, he had made his mark. It'd be another year before prospectors started to arrive in large numbers, but the find he and his partners struck the spark that was to explode as the Cariboo Gold Rush.

It's no surprise that the Cariboo region got its name from the big animals found in great numbers there. The misspelling is attributed to the poor literacy skills of the first miners who came into the area. Initially, only the Cariboo River that fed into the Quesnel River and the Cariboo Lake from which it flowed, had the name. But because gold was discovered along the river's tributaries, the name came to

Named after the Lillooet Natives indigenous to the area at one time, this small town along the Fraser River was of crucial importance to thousands of prospectors hoping to get at the gold deposits in the Cariboo region. The northern terminal of the Harrison-Lillooet Trail, Lillooet was "Mile 0" on the Cariboo Road—the starting point on the long, arduous journey into the Cariboo gold fields. At its prime, Lillooet was one of the largest North American cities west of Chicago, surpassed in population only by San Francisco. Its growth began when gold was discovered in the Quesnel River area. Word quickly got out and miners rushed upriver any way they could. Conscious of the fact that a serviceable road for the prospectors legitimized the crown's rule in the region, Douglas commissioned the Royal Engineers to begin working on the road in the winter of 1858.

include all of the expanding region. Although the Cariboo had no set boundary, at the height of the '62 Gold Rush, a rough approximation of the region was the area encircled by the Quesnel River to the west, Clinton (the old 47 Mile House) to the south, Old Antler Town to the east and Cameronton to the north. Compared with the Fraser River Gold Rush, the Cariboo was pretty far inland. About 30 miles past Lillooet on the Fraser River, Kelly Creek forked off to the northeast. Nearly 20 miles northeast of the end of Kelly Creek was 47 Mile House (Clinton), so called because of its distance from Lillooet. Another hundred miles north were places like Soda Creek and Williams Lake. The Quesnel River forked from the Fraser some 30 miles on and led to Barkerville and Old Antler Town, all places that would enjoy brief flames of fame during the Cariboo Gold Rush.

There probably wasn't a prospector mining on the lower Fraser River during the rush of '58 who didn't, at some point, turn his eyes up river and wonder. Whether he stole a quick glance while manipulating the rocker or whether his gaze was long and fixed, taken while he smoked a pipe and relaxed after a long morning of shoveling gravel on a sluice box, prospectors took to contemplating about the source of the gold they were mining, and dreaming about discovering it. Clearly, if gold was to be found on the lower Fraser, it had to have been carried down the river from somewhere.

In the late 1850s and into 1861 most miners didn't do more than contemplate, and with good reason. Little was known about the territory to the north and northeast because access to the region was difficult. Governor James Douglas didn't even have reliable, first-hand information on the area until 1860. Since the Harrison-Lillooet wagon route had been completed, it was easy enough to get to

Numerous roadhouses were located along the route to the Cariboo, often named according to their distance from larger centers such as Lytton, Lillooet and Port Douglas. There was at least one establishment every 10 to 15 miles, ensuring that those who traveled the route were well provided for on their treks to the northern gold fields. Not that these roadhouses offered anything resembling luxury. Patrons would often pay high prices to drink dubious spirits and sleep on a dirt floor in a ramshackle cabin. Landowners considered themselves blessed if the Cariboo Road passed their property, because there was much money to be made from hungry and thirsty travelers making their way north. The surveyors of the Cariboo Road were powerful men, able to dictate who would reap the benefits of the traffic and who would not. There were charges of corruption among road contractors who conveniently arranged for the road to pass by establishments that they were operating.

Lillooet, but there wasn't much of a white presence in the interior to the northeast of that as late as 1860. There were some 1000 miners at the Quesnel River about 300 miles north of Fort Yale, but relatively prosperous diggings ($10 to $60 a day) and want for a reliable route kept them from venturing inland. A few folks knew of some old Native trails, but the first passage of any significance wasn't built until 1860 when Davidson's Trail linked William's Lake with the bottom of Deep Creek (still a good 50 to 60 miles south—as the crow flies—of the rich gold deposits in the Cariboo). In 1862 the door to the region was finally thrown fully open when Wright's Road was constructed.

As in all gold rushes, there were those whose passion for adventure and the next big strike wouldn't allow them to sit and contemplate. A wilderness not yet carved out was no obstacle to the determined, so by the late 1850s, a handful of prospectors had set out to discover exactly what treasures remained hidden in the remote interior. By 1860, news of their discoveries was to set British Columbia's second golden boom in motion.

Peter Dunlevy and his partners first found gold in the southern reaches of the Cariboo in the spring of 1860. Their discovery attracted a few more miners. The initial find was followed by a more substantial find a few months later. Late in 1860 two prospectors in Quesnel decided they had wondered long enough about the source of the gold they were digging. With the best of the mining season long over, Doc Keithley and George Weaver set off towards the mountains that fed the Quesnel River. The journey brought them into the heart of Cariboo country, where the terror of Fraser River white water met its match in the intimidating, jagged reaches of the upper Columbia Mountain range. A miner was later to write of the place,

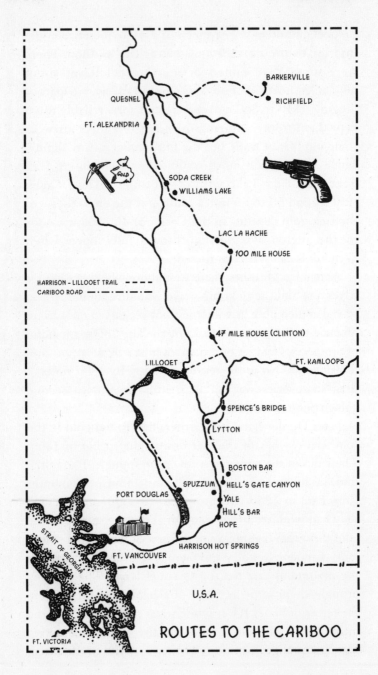

"It seemed as difficult to access as the Arctic regions." He wasn't far off. Just as men were driven to explore the waterways of British North America's frozen north in search of a passage that might give easy access to the Orient's riches, so were men driven to the Cariboo in search of their own fortunes.

Panning from stream to stream, Keithley and a fellow named Diller (Weaver had gone off on his own) kept moving inland until they reached the Cariboo River, a tributary of the Quesnel River. Conditions were such that they couldn't pan on its shores, so they built a chute to direct the water to where they could use a rocker. They hit it big, making about $100 per bucket of gravel. Though it was a substantial find, it wasn't enough for Keithley. He noticed that the gold had a scaly appearance and experience suggested that such gold was washed from a mountainside. He continued on into the interior while others flooded to the strike off the Cariboo River on what folks had come to call Keithley Creek. Soon miners were taking $25 to $50 a day, a handsome return. As the creek was staked, prospectors took to nearby streams. A fellow named Harvey did so well that a creek was named after him as well. Before the snow fell, his claim had netted $25,000 for him and his partners.

By 1860, prospectors were using the name Cariboo to identify the region encompassed by the Cariboo River's tributaries. The region continued to expand, thanks again largely to the efforts of Keithley. In September, he set out with his old pal George Weaver along with John Rose and Ben McDonald in search of the source that was feeding the Cariboo diggings. They headed north and every step was uphill through territory that was so dense and inhospitable that even the animals avoided it. Their traveling was further impeded by the shortening days of fall and their heavy, bulky miners' packs. Eventually they ended up on the western edge of a plateau (Snowshoe Plateau, between Old Antler Town and Keithley Town) that gave an awesome view of the land and rivers to the south and west. Majestic as it was, the view was nothing compared to what they would soon see.

Stumbling upon a creek that flowed through a narrow rock canyon, the men threw down their gear for a closer look. The creek sparkled! They rubbed their eyes. It had to be the sun's rays catching the gentle ripples on the water. They moved closer. It wasn't the sun. The shimmering light was coming from the riverbed that sparkled like the starry night sky! Rose dipped in his pan. A little shaking revealed $75 worth of gold. McDonald did the same. $100! What a find! The stream was named Antler Creek, and it cut a path about five miles east of soon-to-be Barkerville. They even found sunburn gold, deposits exposed to the sun that had begun to oxidize. None had ever been involved in such a wealthy discovery and that was hardly surprising, since it was unmatched in British Columbia.

The four prospectors worked for a few days until the snow came and then they had to halt operations. They built a shelter for the winter; they'd brave the worst of what that season might bring to ensure they could start mining as soon as possible in the spring. However, to make it through the winter they needed supplies, so some of the party headed back to Keithley Town. Wizened old prospectors, who noted that the party was fewer in number, smelled a strike. When Rose and McDonald made their way back to Antler Creek, they were followed. The news of the find spread from there. Men straggled in on snowshoes through the winter, and snow that was often ten feet deep (on the level ground, never mind the drifts) presented no significant obstacle. By spring, the shores of the stream were fully staked and crowded with hundreds of dug-outs hurriedly built by those who had chosen to abandon their Quesnel claims and, like the Keithley party, to brave the winter for first crack at spring mining. May 1861 saw snow still on the ground but many of the 1200 men in the region were already busy, some taking out as much as $300 a day.

Dutch Bill Dietz wasn't one of them. He stood on the rocky cliff that served as one of the canyon walls on Antler Creek and cursed. When he stopped to catch his breath, he bent over and picked up his shovel. He raised it above his head and was about to throw it into the stream. At the last second he threw it on the ground instead. Then he collapsed, joining it there.

"Dutch Bill," he laughed. "They might as well call me Bad Luck Bill. Wouldn't be any less true."

William Dietz did know about bad luck. Back in '58 he'd joined the flow of California miners bound for the Fraser and fortune. Dietz found the river alright, but whether he swirled a pan or shook a rocker, he was never blessed with anything more than a little color. The most he had discovered was just enough to allow him to remain in gold country. But he was never in a position to pop the cork off a champagne bottle. Hell, any place that even sold champagne was out of his league. No, it was a hand-to-mouth existence for Dutch Bill. And then things took a turn for the worse. One night in Lillooet, while enjoying a rare night at a place that sported a whip-sawn plank over two barrels and called itself a saloon, Dietz discovered that his gear had been stolen. He couldn't even enjoy tying one on.

Like many of the miners who labored on the Fraser in the late 1850s, Dietz had slowly worked his way upriver. He was up on the Quesnel, panning a half dozen dollars a day when he heard news of the Antler Creek strike. Dietz wasted no time in heading for what folks were calling the new El Dorado, but he arrived to find the creek solidly staked. As he sat near the canyon's edge, he wondered why it should be any different. It was the story of his life, a day late and a future of cleaning up the leavings at best. Dietz stood up. He'd have none of that this time. It was spring. He had enough supplies to last to summer at least. He'd give it another shot.

Dietz joined up with a couple of fellows. They traveled north and west along Snowshoe Plateau until they had pretty well reached its western edge. Night was falling when they almost stumbled off the plateau into a valley whose crevasse was hidden from view. The trio slipped down the mile drop that led to the creek at the bottom of the valley, made camp and called it a night. They spent the following day prospecting, but the best any of them came up with was a pan bearing about $.30 worth of gold. Discouraged, they sat down to another meal of beans. Everyone but Dutch Bill figured he'd had enough of prospecting for the day, so when he went out for one last try, he went alone. Moving upriver, he finally fell upon a spot that yielded $1 a pan, but that was as good as it got. He returned to tell his friends of the find. They weren't impressed.

"You know what, Dutch Bill? You can have this danged creek for yourself," said one of them.

"Sure. We'll name it after you."

"You do that and I'll open for you the first case of champagne that comes into this country."

"Hell, Bill, that's a deal."

"Don't matter if we call it Williams Creek," reflected the other. "Everyone'll know it as Humbug Creek. All they'll find here is backache."

His companions headed back to Antler Creek. Word spread that there was gold on Williams Creek. Everyone knew that it wasn't much, but there was no sense hanging around waiting for someone to abandon or sell his claim. The first of the miners arrived in late winter but it was

The Cariboo

a few months before they could start mining in earnest. The shores were overgrown with brush and trees that had to be cleared. Rockers and sluices had to be built. And as the first rains gave way to summer's heat, they were confronted with the biggest impediments, blackflies and mosquitoes. The physical labor of prospecting was darn near pleasurable compared with the attack of the insects. Perhaps they could have been more easily endured had someone actually found some gold, but no one was turning up more than Dietz and his partners had. All was to change on one summer day, when a fellow named Abbott took to thinking about a layer of blue clay he had hit some eight feet below the surface. At first he thought it might be bedrock, but he decided to try and break through it. Good thing he did. Within two days he had 50 ounces of smooth nuggets!

Men abandoned their pans and rockers and sluices and took to digging holes. Miners were finding two or three hundred ounces of gold a day; at the Steele claim they took out more than 400 ounces! The Lowhee claim, 400 feet long and 12 feet wide right alongside the creek, mined over 3000 ounces in three months. Abbott was doing so well that he began measuring his daily take in pounds and it wasn't rare to hear that he had found 30 pounds on a given day. Within a few months he had a profit of more than $80,000. Most of these big claims were worked by paid labor but they had to be given a decent wage to prevent them from working their own diggings. Dutch Bill was there for the ride but he could only watch as miners pulled up buckets of gold to his buckets of gravel. Sure he made some money, pretty well everyone on Williams Creek in the summer of '61 did, but what he gained in fortune he lost in health. He picked up some kind of disease in the Cariboo and could never shake it. For a time, however, his name was associated with the biggest find in the Cariboo.

Pictured above is one of the gold prospectors who made the trek up to the creek named after William Dietz. Washing gravel dug from Williams Creek, he is employing a prospector's tool called a rocker to separate the ore from the dirt. The rocker was used for placer mining when there was not enough water available to make good use of a sluice box. While a lone prospector could work a rocker, it was most efficient if three operated it. One man dug the gravel, one brought it to the rocker, and one poured water and rocked it. Allowing at least five yards of gravel to be washed per day, the rocker was a more efficient machine than the sluice. Miners who knew what they were doing could wash about 200 buckets of gravel in one day. Some of the biggest strikes in the Cariboo were made along Williams Creek, with hauls as large as 30 pounds of gold per day.

Antler Creek and Williams Creek were the finds that set the Cariboo Gold Rush into motion, but it didn't happen overnight. Throughout 1861 most of the miners who came to Williams Creek were already prospecting on the fringes of the Cariboo. It wasn't until they began to filter down to Yale, New Westminster and Victoria in late fall and early winter that news of the strike spread. And when it did spread, it was like wildfire. Local papers stoked the flames. Throughout October 1861, the *Victoria Colonist* underscored the scheduled arrivals of the *Otter* (the steamship that regularly plied the New Westminster-to-Victoria route) with a stream of mouth-drying reports, such as, there were 45 passengers and $100,000 in gold dust; 130 passengers and $150,000 in gold dust; 72 passengers and $250,000 in gold dust; 2 miners with $20,000 each. It was the greatest gold shipment to Victoria, *ever*! By the end of 1861, the Cariboo had produced a good $2,000,000 in gold. The rush was about to become a torrent.

James Douglas, Governor of British Columbia since 1858, was always one to be prepared. His experience with the Fraser Gold Rush had only strengthened his resolve when it came to ensuring that the Cariboo Gold Rush would be a civil affair. Douglas' guiding principle continued to be that Britain retain sovereignty over the gold country, which he believed could be achieved through the creation of laws and the presence of justice officials inclined to impose them. Judge Matthew Baillie Begbie (the top colonial magistrate) and a handful of regional Gold Commissioners (who also served as justices of the peace) were, for the most part, well suited to that task. But Douglas was not convinced that law alone was adequate to ensure sovereignty. He considered that it was essential to solidly link the mainland colony with Vancouver Island. The link was forged with a transportation system that funneled miners in and out of the Cariboo through Victoria. The fact that British Columbia remained British was a testament to Douglas' determination as well as to the skill and efforts of the Royal Engineers and others who built the transportation routes.

By late 1860 gold commissioners following the flow of miners were beginning to penetrate the Cariboo. They began to send Douglas some of the first official reports about the region and what they wrote was hardly favorable. H.S. Palmer of the Royal Engineers, who was charged with making some initial surveys of the territory, confirmed the frightful evaluations, "It is difficult to find language to express in adequate terms the utter vileness of the trails of the Cariboo, dreaded alike by all classes of travelers; slippery, precipitous ascents and descents, fallen logs, overhanging branches, roots, rocks, swamps, turbid pools and miles of deep mud…. The only good parts are on the actual summits of the bald hills; even the upper portions of the slopes are, in many places, green, spongy swamps, the headwaters of the radiating creeks; and, directly the forest is entered, the more serious evils begin."

While such conditions brought many a curse to the lips of northern-bound miners, a second unfortunate effect was to cause the prices of goods to soar. From a miner's perspective, the whole problem was well summed up in a letter by one of their own, Radcliffe Quinn. "I tell you it is a hard road to travel. You have to carry your own blankets and food for over three hundred miles and take the soft side of the road for your lodgings and at daylight get up and shake the dust off your blankets and cook your own food for the day and take the road again. When you get in the mines you have to pay up to a dollar a pound for everything you eat as it has to be carried on mules and horses on their backs with a pack saddle."

Douglas got a firsthand view of the region in the summer of 1861, when he traveled up to Barkerville. His journey highlighted two important points. First, there was much more gold being mined than was actually making its way back through Victoria and second, a road into the heart of the Cariboo was necessary. The two points were connected because the economic benefits of having the gold travel through the island capital were numerous; few businessmen didn't know that the metal burned holes in miners' pockets. Douglas harbored no illusions about the difficulty of the task, but he had an ace in the hole, one he was certain was up to the job, the Royal Engineers.

An elite and highly competent branch of the British military, the Royal Engineers (sappers) had already proven their usefulness during the rush of '58 when they had shaped the Harrison-Lillooet trail into a serviceable wagon road. However, because there were so many wagon-boat transfers on the Harrison-Lillooet road, it proved too costly for freighting goods upriver. So, in early 1862 Douglas commissioned the Royal Engineers to survey wagon roads along the Fraser River from Yale to Boston Bar and along the Thompson River from Lytton to Cook's Ferry (Spence's Bridge). The assignments were a challenge. The road had to be 18 feet wide and there were many places where even the best route demanded that solid granite be blasted to give access. To make the route effective, the Fraser had to be crossed. The sappers surveyed a site at Spuzzum, 12 miles north of Yale. By September 1863, Joseph Trutch and his contractors (Halliday and Company of San Francisco) were putting the finishing touches on the $45,000 Alexander Suspension Bridge. Trutch was granted a five-year toll as payment. Onlookers held their breath when the first load was transported across; the bridge sagged a mere quarter of an inch. The way was finally clear for Douglas' most ambitious project, a road to Barkerville.

The Royal Engineers surveyed the route and built the first six miles, finishing in the summer of 1863. Contractors built much of the remainder of the route, with entrepreneurs including Gustavus Wright, William Hood and J.C. Callbreath prominent among their number. By 1864, a miner could start at Lytton, at the fork of the Fraser and Thompson Rivers, and make his way up to Quesnel, at the fork of the Fraser and Quesnel Rivers. For the most part, travelers were well east of the Fraser, because the Cariboo Road took them through Clinton and Lac la Hache in the interior. Just north of Lac la Hache, the route veered west again, joining up with the Fraser at Fort Alexandria. From there, a miner could take a steamer to Quesnel (eventually the sappers built a road connecting the two). By 1865, a road covering the last 43 miles east to Barkerville was finally constructed.

The rush of '62 had begun to peter out by then, but even during construction the road had a great impact on activities in the Cariboo. The cost of freighting fell dramatically, on average from $.75 to $.15 a pound. Even with the tolls charged at various places along the route, it was no longer a dollar a pound for everything a miner ate, as lamented by Quinne. Flour, once $2 a pound, could be had for $.35, while potatoes, once $1.15 a pound, dropped to $.25. Butter was no longer a luxury at $5 a pound, but a mere $1.25. The prices of just about everything decreased, often by as much as half and sometimes more. In the heady days of the rush, when miners were paid a daily wage of tens of dollars, even many of those who didn't strike it rich could finally escape a pauper's life.

The Cariboo Road also brought new wealth to folks who had given up on mining. Since the days of the rush of '58,

Running 400 miles along the Fraser River Canyon between Yale and Barkerville, the Cariboo Road was built as a wagon route to the gold fields of the Cariboo region. Governor James Douglas conceived the idea of the road. He believed that a well-constructed transportation route into British Columbia's interior would increase his authority over the region and ensure that the colony remained under British control. The road also made the Cariboo more accessible to miners, reducing their transportation and supply costs, and thus making them less likely to cause trouble. In 1862 the Royal Engineers surveyed the road and completed work on the route's two most difficult stretches: the 6 miles from Yale to Boston Bar and the 9 miles from Cook's Ferry along the Thompson River. The remainder of the road was completed by private contractors in 1865. Because the proposed route was often along precipices or blocked by solid rock that had to be blasted, it cost more than $1.25 million.

entrepreneurs had built roadhouses (sometimes called restaurants) along the northern trails. The pattern had changed little in the rush of '62. A glance at a map of the Cariboo Road shows that it was peppered with roadhouses usually separated by 10 to 15 miles, but often even less. Many roadhouses were labeled Mile House, the number of miles reflecting the establishment's distance from Lillooet. Occasionally, the roadhouses were named after the owner, although maps didn't come out fast enough to keep up with the changes in ownership. In terms of earning a living, running a roadhouse had about as much security as mining. Dishonest money lenders, unpaid bills, and foreclosures were the humbug creek of the service industry. Others had the misfortune of seeing the Cariboo Road constructed some distance from their establishment. Folks who found themselves in such a position couldn't blame dumb luck. The fact was that road contractors made a good profit operating roadhouses, so the choice to build a road where there had never been a trail (and where, therefore, there was no competition) was always desirable.

Generally roadhouses were built in open areas, with good access to water and arable land for growing vegetables. Although the quality of the establishments improved with time, in the early days of the rush it was rare to stumble upon one that wasn't a hastily built log structure graced within by little more than a stone fireplace and a bar. With a door but no guarantee of a window, they were stale, stuffy and stinky affairs. There was nothing fancy about them, but amenities in the bush were relative. Just finding a place to have a half decent meal, a place to sleep, a drink or three, and a game of cards, was akin to finding the leprechaun's pot for many a tired, lonely miner. In the early days of the rush, prices for services at a roadhouse were high, reflecting what the market would bear, but eventually they leveled out at $.50 a meal up to Quesnel and $.75 at points beyond. It hardly mattered that a miner would be following up the evening's activities by sleeping in his blanket on the roadhouse floor, usually getting a case of lice for his troubles. Likely he didn't even take offence at the barkeep's sleeping on top of the bar to protect his stock of liquor, although many a miner probably wondered how solidly he slept.

Most nights he probably slept quite well, but there were occasions when rest was fitful at best, interrupted by images of violence. The Cariboo Road wasn't fussy about the morality of the folks who traveled on it and the presence of criminals along the route and in the roadhouses that dotted it was inevitable. Where there was gold, there was bound to be theft and, because miners weren't anxious to part with their hard-earned profits, most robberies were accompanied by bloodshed. The Gold Escort emerged as one solution to the problem. The Gold Escort was the brainchild of Phillip Nind, the Alexandria Gold Commissioner. Formed in 1861, the Gold Escort consisted of twelve uniformed, armed and mounted men. They promised to get a miner's gold to Victoria at a cost of 1 shilling per ounce. The charge seemed excessive to many money-pinching miners, so the Gold Escort ceased operations within a few months. It was re-established in 1862, but banks mostly used its service. Perhaps as a result of their miserly ways there was more than one story of a miner losing his poke and occasionally his life.

While murderers grabbed the headlines, the most significant criminal element in the Cariboo was the professional gambler. As Judge Matthew Baillie Begbie noted, where there was gold to be found and liquor sold, professional gamblers were sure to follow, "as the carrion crow scents the dead on a battlefield." The gamblers caused no

While James Douglas made it his top priority to ensure law and order prevailed in the Cariboo, there were few prospectors who believed their golden gains were adequately protected from a governor hundreds of miles away. Many miners got restless as the number of undesirables increased along the Cariboo Road. From Lillooet to Barkerville, predators plied their trade behind a hand of cards at a gambling table, or waited along the road with loaded weapons. Pictured above is one man's solution to the troubles that plagued prospectors who were trying to get their gold back to Victoria. The Gold Escort, formed by Alexandria Gold Commissioner Phillip Nind, provided an armed escort for any gold shipments from the Cariboo country to Vancouver Island. The four-man guard above is about to depart from gold-rich Barkerville.

end of problems and they were hard pressed to find a friend from Clinton to Barkerville. James Anderson was a Scottish immigrant who became known as "the bard of the Cariboo." In addition to his many songs, which enjoyed popularity in the region, he wrote a series of Sawney's Letters that were published in the local paper. His displeasure regarding the presence of the professional gambler summed up the feelings of many an honest miner.

> *There is a set o' men up here,*
> *Wha never work thro' a' the year,*
> *A kind o' serpents, crawlin' snakes,*
> *They fleece the miner o' his stakes;*
> *They're gamblers—honest men some say,*
> *Tho' it's quite fair to cheat in play—*
> *IF IT'S NO KENT O'—I ne'er met*
> *An honest man a gambler yet!*
> *O, were I Judge in Cariboo,*
> *I'd see the laws were carried thro',*
> *I'd hae the cairds o' every pack*
> *tied up into a gunny sack*
> *Wi' a' the gamblers chained thegither,*
> *And banished frae the creek forever.*

It wasn't just that the professional gamblers fleeced the miners; they also caused all kinds of other problems, as events at roadhouses illustrate.

During the fall of 1862, in a dark corner of the cavernous Menefee's Mission Creek Roadhouse, a handful of miners were playing Spanish monte. The stakes were small, as few had much to spare at this time of year. No one paid attention when the door to Menefee's was thrown open and a blast of cool air shot into the room, but heads popped up when a $20 gold coin was thrown on the table immediately thereafter. In a reply to the ante, one of the players drew his revolver and the man who had thrown down the coin did the same. The stranger fired first and his bullet thudded into a log. When the other man fired, his arm was jarred at a crucial moment and his bullet went astray, killing an innocent bystander. The stranger took the opportunity to flee. Eventually the truth came out. The man playing cards with the miners was Gilchrist, a professional gambler from San Francisco. While down there he had made enemies with the stranger over a game of cards. They pledged to kill each other upon next meeting. For being in the wrong place at the wrong time, Gilchrist went to jail. The stranger was never seen again.

The details were somewhat different at an incident in Maloney's Roadhouse near Williams Creek but like the tragedy at Menefee's, the episode involved a professional gambler and a murder. One day in the summer of 1862, Reverend Dundas, cleric for the regional Anglican bishop, stopped into Maloney's for breakfast. As he ate, he was joined at his table by a miner. Dundas was an outgoing man, by nature and profession inclined to strike up a conversation with those he encountered. Try as he might with this fellow, however, the only response he got were jabbing dark glances from beneath an intimidating single eyebrow. Dundas shrugged and went back to his meal. Finished and about to stand up, he was frozen still as a man bolted into the roadhouse. "Look out!" he shouted. The warning wasn't out of his mouth when the man Dundas had been eating with darted out of the roadhouse. While the men in the Maloney's began to discuss the odd turn of events, two constables from nearby Richfield showed up. It turned out that the fleeing man was "Liverpool Jack," a professional gambler who had killed a man the night before in Richfield. There's no record as to whether Dundas dipped into the ceremonial wine to ease his frayed nerves.

There were a handful of miners in the Cariboo who had places named after them, folks like William Deitz and Doc Keithley. But only William Barker could say that his namesake was the biggest gold rush boomtown ever. Though Barker dreamed of making the big strike, as did every miner, even he would have laughed at the mention of such a possibility. Billy had no idea what was in store when he first started to dig a ditch in an unstaked area above Williams Creek, no notion that'd he be the midwife of a place that was said by contemporaries to be the biggest town west of Chicago and north of San Francisco. Barker's discovery was to transform the northern Cariboo.

Barker's story was, for the most part, about as common a tale as a fellow could find in gold country (all that really changed from miner to miner were small details like names and places). Born in England in the early 1800s, his early years saw him employed as a riverman, working the barges in Cambridgeshire. The railway boom of mid-century meant that there wasn't much of a future for a riverman and Barker made his way west, while his wife and daughter stayed in England. By the late 1840s, he found himself in California searching for gold. When that rush played out, he headed north for the next El Dorado and was on the Fraser River by 1859. Perhaps it was just as well that he had a good case of gold fever. There was no reason to go back to England because his wife had died.

Barker spent much of the rush of '58 around Canada Bar, but he never made more than he needed to eke out a living. When the news of the find at Williams Creek floated down river, he packed up with a few friends and made tracks north determined to cash in on the bonanza. When they arrived in the spring of 1861, there was nothing left to stake near the

William Barker, a bargeman from eastern England, can truly take credit for the gold rush boomtown of Barkerville that is named for him. It sprang up practically overnight around his claim and those of the others who threw in their lot with him. This man, who persevered in the face of impossible odds and with the financial backing of Judge Begbie, was instrumental in changing the economic face of B.C.'s wild and isolated interior. It wasn't isolated for long because, once news of the size of Barker's strike made the headlines, thousands of prospectors abandoned their claims on the Fraser and headed north to stake out the creeks surrounding Barkerville. Although Barker made a fortune on his claim, he died a poor man at the Old Man's Home in Victoria.

gold find on the creek, so the men continued north, eventually staking a claim in Mink Gulch near Richfield. All they found at the Barker and Company claim was a whole lot of gravel. Discouraged, Barker and his associates decided to take a chance farther downstream. Local miners considered the area barren but it was one of the few places left that was both easily accessible and unstaked and that was enough for Barker. The claim was registered in August 1862. Barker and Company consisted of seven men by this time so that gave them the right to stake 700 feet of channel.

 Before they started digging, Barker and his companions built a log-walled shaft house that served as shelter and a base for operations. The building wasn't up before miners started dropping in to see what the crazy Englishman was up to. They were merciless in poking fun at the small group of men, so certain were they that Barker and his friends were wasting their time and efforts. The wisecracks only served to steel Barker's determination. Within a week, the eight men (one more had joined the company) turned their efforts to sinking shafts. The first two shafts brought nothing but backache. The third shaft reached about 35 feet, already substantially deeper than anything dug around Williams Creek. Then their money ran out. Barker wasn't about to give up, so in desperation he searched the region for someone to bankroll the operations. His salvation appeared in a most unlikely form, a six-foot man with a waxed moustache and a Vandyke beard, Judge Matthew Baillie Begbie. Begbie wasn't much interested in investment, though charges of financial misconduct and conflict of interest were leveled at him. He later revealed that he gave Barker funds because it was cheaper than removing the destitute men from the region.

 The men went back to their shaft. At 50 feet, even with the money given by Begbie, the men felt like giving up.

Tradition has it that only the Barker's stubbornness kept the men digging. Whatever spurred them on, it wasn't long before despair turned to exhilaration. The gravel started to sparkle. They had struck the "head of the lead," the underground channel that led to Williams Creek. Within moments of the first strike, they were making $1000 a foot. Ten hours later they had collected 124 ounces of gold that, when divided up, gave each man $20,000. The gibes directed at Barker and his associates were heard no more and there wasn't a miner who wasn't anxious to be a neighbor to Barker's rough old log structure. Like mushrooms after a good rain, cabins popped up around Barker's claim. From the beginning, the collection of buildings was known as Barkerville.

The discovery of gold was enough to ensure the growth of Barkerville; however, a decision by the Royal Engineers in 1863 was to give the town some long-term security. When the sappers were detailed to survey the northern reaches of Cariboo Road, they were met with some pressure from folks living in Quesnel Forks to use their town as the center for distribution of goods into the gold fields. The problem was that a mountain range divided Quesnel Forks from places like Antler and Richfield. Lieutenant Palmer of the Royal Engineers was responsible for making the decision that would shape Barkerville's future. The Cariboo Road would hug the Fraser River to the mouth of the Quesnel River and stop at Quesnel. It was expected that Quesnel would serve as the supply center for the whole region. The *Victoria Colonist* proclaimed that the town was sure to become the largest in the British Columbia interior. It wasn't to evolve that way. With the discovery of gold in Barkerville, the Cariboo Road was extended inland to that location. Since Barkerville was on Williams Creek and closer to the other gold strikes, it quickly took on the supply responsibilities that Quesnel expected to have.

Barkerville's population exploded and, within a year, it had grown to more than 10,000. Many of them lived along Main Street, Barkerville's only road in the early years. It wound its way through the valley bottom, in some places hugging Williams Creek, while in other places redirecting the water's flow with wing dams so as to prevent the town from flooding and to allow for new mining sites. Try as they might, however, residents found it impossible to prevent Main Street from turning into a wet, muddy mess especially with spring run-off from the adjacent hills. To solve the problem, few structures were actually built at ground level. Most were supported by some sort of trestles. As a result, entryways were well off the ground. Boardwalks ranging from three to nearly six feet off the ground and lining both sides of Main Street gave access to buildings. Neither was there much attention paid to aesthetic details like ensuring boardwalk planks were of uniform lengths. The uneven boardwalks along with the many hastily constructed buildings that were laid out without the least of planning gave Barkerville a rough and raw appearance. Certainly no one would confuse it with the more refined towns of the east.

The demand for wood to be used for construction was insatiable. Not only was it used in the town itself, but wood was important for shoring up shafts and wing dams and for building sluices and water-wheels, all of which were common sights in and around town. A most visible result of the insatiable demand for wood were the barren hillsides. Before 1862 the trees were so thick that it was a challenge for a prospector to make his way through them. Within a few years, only the stumps remained, giving the hills an uncanny resemblance to the stubbly cheeks of an unshaven miner. The look of the place wasn't much improved by the piles of tailings (the material removed from shafts) that quickly filled the creek bed.

The town of Barkerville, shown here with Williams Creek in the foreground, was located in the heart of the Cariboo. At its roaring peak of 10,000 gold-grubbing souls, the town marked the high tide of the Cariboo Gold Rush. Counted among the liveliest of the gold boomtowns, it was in the middle of a mountainous nowhere, sprouting around the claims of the fabulously successful but ultimately doomed Billy Barker. Over the course of one year, miners' crude cabins and canvas tents were replaced by wooden establishments built from the rapidly thinning forest surrounding the town. There were saloons, general stores, dancehalls and gambling houses, all of them packed with grizzled miners buying services at prices that were far too high. During its early years, supplies were scarce, arriving with slowly moving pack trains or on the backs of miners. That all changed when the Cariboo Road reached Barkerville.

Of course, those attracted to Barkerville weren't much interested in how the place looked. They came for the gold and the opportunity to have a good time spending it. Barkerville obliged. The first services were the basics—stores, hotels, livery stables, feed shops and restaurants. There was nothing fancy about any of these. The early hotels, for example, weren't much more than bunkhouses. But as the town boomed and competition between businessmen mounted, services improved. Restaurants advertised the quality of their cooks and hotels noted the modern style of their furnishings and the comfort they provided. Within a few years there were amenities usually found in only the most mature of communities. Civic-minded citizens collected money for a hospital. Those who had more cultured tastes frequented the library or the weekly show given by the Cariboo Amateur Dramatic Association at the playhouse. Spiritual needs were also met by the active Catholic, Anglican and Methodist denominations in the town. And if a fellow wanted to keep abreast of the news, he read the local paper, the *Cariboo Sentinel*, instead of relying on gossip.

For those folks who preferred a kick to their leisure time activities, Barkerville had much to offer. There were plenty of saloons, ranging from quiet, intimate places to loud, cavernous ones. At first, they offered little more than alcohol, but it wasn't long before they were livened up with pianos, billiard tables and dancing girls. Miners who didn't like saloons could purchase their booze from the Barkerville Brewery, which produced a prize-winning Triple X Ale. However, most folks who bought it probably didn't get to drink the ale at its best, since Triple X was so popular that it rarely had time to properly ferment before it was sold. Another drug of choice was opium, legal until 1908. It was often easier to find a gaming den or brothel in Barkerville than it was to find gold, which is saying something because there was plenty of gold to be found on Williams Creek from 1862 to 1863.

In 1868 the town of Barkerville burned to the ground. Legend has it that the fire was caused by a miner trying to steal a kiss from a washer woman. He was rebuffed in his attempt and, as he staggered backward, he knocked a pipe from a stove. The hot pipe came into contact with some canvas and flames erupted. The fire spread quickly through the wooden town and, by the next day, there was nothing left but charred embers. By then the gold rush was over and Barkerville was more than a gold town. Folks looked upon it as home and they vigorously took to rebuilding it and getting on with their lives.

Billy Barker wasn't so fortunate. He mined through 1862 and, with the coming of new year, he left for Victoria. While there he met and married Elizabeth Collyer. His new bride wasn't one for a quiet life at home. Rather, it was the high life for her and so the couple spent night after night in saloons and restaurants. It was a great way to enjoy life but also a costly one. To meet the expense, Barker and his wife returned to Barkerville in the spring of '63. It didn't take him long to fill his pockets. Again they left the Cariboo and discovered that it took even less time to spend it all. Barker returned one more time, in the summer of '64, but his claim was played out. Poor investments and bad luck left Barker with about as much money as he had prior to his strike on Williams Creek. His wife abandoned him in the spring of '65. Although Barker was to make one more strike on Poor Man's Creek near Beaver Pass, he was destitute when he died in 1894 at the Old Men's Home in Victoria. There were few miners who struck it as rich as Billy Barker (it's said that he took half a million dollars from his Williams Creek claim), but among a group that mostly chose to live for the moment, his ending was hardly unique.

While Barkerville was certainly no San Francisco, it was beginning to evolve past its primitive origins by 1868. Services were no longer limited to the saloons, dancehalls and gambling houses that had been erected during the town's first months. Local churches provided pious miners with weekly services, a group of cultured residents established the Cariboo Literary Institute and the Theatre Royal was a regular venue for stage shows. A number of stage companies also established terminals in Barkerville, providing a closer link to Vancouver Island. The one thing that all these establishments had in common, from gambling house to house of worship, was that they were built of wood from the surrounding forests. At the end of the particularly dry summer of 1868, disaster struck. A roaring fire swept through town, destroying almost every building. Though the above photograph is testimony to the scope of the destruction, the citizens of Barkerville banded together and, within a remarkable six weeks, rebuilt most of the town.

It was a rare Saturday night indeed when Barkerville wasn't loud and busy with miners drinking away their frustrations…or successes, but on this night the town was nearly empty. The Fashion, Martin's, the Crystal Palace and all the other big name saloons in town might as well have hung closed signs on their doors. In fact, business was pretty slow in towns as distant as Stanley and Old Antler because men from every corner of the creek had chosen to gather at James Loring's Terpsychorean Saloon in Cameronton, just north of Barkerville. And with good reason. Loring had trumped them all; he had the hurdy gurdy girls.

There were many folks swept along in the rush of '62, as there were in every gold rush, who had no intention of mining. Mostly, such people were frowned upon by prospectors, as leeches, out to make money from their hard work. But no one complained about the hurdy gurdy girls, or gurdies as they were usually called. The gurdies were German dance hall women brought to the Cariboo (perhaps as early as 1863, but no later than 1864) via California, where they had been popular. Most agree that the dancers got their name from the hurdy-gurdy, a popular musical instrument in rural Europe. It was played with buttons and a wheel that rubs a string. Others joked that hurdy was an old English word for buttocks or hips, so it was appropriate name for the ladies given the type of dancing they did.

Of course, the miners at the Terpsychorean didn't give two figs about the origin of the name; they just wanted to dance. The arrival of the women was much anticipated and appetites were whetted with advertisements that had regularly appeared in the *Cariboo Sentinel*.

> # THIS IS THE OPENING NIGHT
> # THE TERPSYCHOREAN SALOON!
>
> ### CAMERONTON JAMES LORING, Prop.
>
> This Magnificent Saloon, which is the finest in the Cariboo, for Terpsychorean Exercise, will be opened THIS (Saturday) EVENING, the 17th June, to the lovers of MUSIC and DANCING.
>
> First Class order will be preserved and the proprietor invites all his friends to give him a call.
>
> The Bar is stocked with the finest Liquors and Segars.

To be sure not to miss any potential customers, Loring also had posters tacked up in the towns along Williams Creek. Anyone who couldn't read stared at the signs until someone who could passed their way. A fellow never tired of reading or hearing that the gurdies were ready to dance. It was the kind of news that spread through the mining camps with all the speed of a gold strike. The gurdies were coming and they only wanted a dollar a twirl! It was little enough. Weeks and months working alongside men who reeked of sweat, amidst sharp, hard rocks gave a miner a certain willingness to part with a dollar in order to smell perfume and feel a soft curve.

"Gentlemen!" came a cry from the stage. Eyes were drawn forward and focused on James Loring. "The hurdy gurdy girls will soon be out and ready to dance."

An enthusiastic cry went up among the men.

"I notice that the place already has a warm glow," Loring noted to continued guffaws. "But to get the feet tapping, our piano player is going to play a few tunes." Gurdies and fiddles weren't good enough for Loring. He had an upright piano shipped in from Victoria (the last

60 miles it was carried by four men) and he was going to get his money's worth out of it.

Cheers turned to groans, but the crowd fell silent when the pianist broke into a rendition of *The Lover's Lament*. By the time he played *Do They Miss Me at Home* the men were rocking too and fro like babies, sniffling and blowing their noses.

"*The Skedaddler!*" someone shouted. "Play *The Skedaddler.*" The man who requested it must've been down on his luck, because the song was about a unsuccessful miner who left the Cariboo as well as his debts and responsibilities. The reception the song got suggested there were many empathetic souls in the place. With the first line, everyone was singing, and the place sounded like a mix of fighting cats and grumpy bears.

> *I'm dead broke*
> *I'm dead broke*
> *so I've nothing to lose*
> *I've the wide world before me, to live where I choose*
> *I'm at home in the wild woods, wherever I be*
> *'Tho dead broke, 'tho dead broke, the Skedaddler is free.*
>
> *'Tho Creditors curse me*
> *I care not a straw*
> *I heed not old Begbie*
> *I laugh at his law*
> *There is game in the mountains, the rivers yield fish,*
> *And for Gold*
> *I can prospect wherever I wish.*

On it went for another four stanzas. When the song ended with the hope that the Skedaddler might yet win honors and wealth, the place went up in a cheer. Even louder were the hoots and hollers that followed when Loring introduced the hurdy gurdy girls.

There were four of them and though more would follow, they danced as quartets. By the measure of the men's faded memories, they were among the prettiest women these miners had seen. What they might have lacked in beauty, they made up in energy and enthusiasm. Their smiles were wide and their cheeks were red. Great flowing skirts were pulled tight around their waists. Miners pushed and shoved to be the first in line to dance. A couple of fights broke out and men laughed as they stepped over the wrestling men who had lost their places. Soon they were spinning on the floor.

The gurdies danced well enough considering the few lessons they'd received and the months of practice in western mining camps. Not that it took much skill to participate in the frolicking that passed for dancing in the mining communities. With the strike of the first note, the gurdy's feet were off the ground as the miner swung her high. Urged on by his companions, who were usually close around the dancers in such a tight circle that a heel in the chops was as likely as not (a welcome badge to serve as a reminder of the evening), the miner strove to swing the gurdy high enough so that she could dance on the ceiling. As described in the *Cariboo Sentinel*, "If you ever saw a ring of bells in motion, you have seen the exact position these young ladies are put through during their dance...." The best male partners were considered to be those who could swing the gurdies the highest. They all gave it a school boy's enthusiastic effort and the strong arms of many a gold digger ensured that the smoke-blackened boards above were well streaked with scuff marks.

The gurdies were also practiced at the art of the sale. Part of their job was to encourage miners to buy drinks.

After a dance, the gurdy joined the miner at the bar, where he was encouraged to buy a whisky for himself and a juice for her. The drink ran another dollar and the gurdy collected half the cost. A gurdy's actual wage was less than that; however, because Loring took a commission from all sales. He did well enough, with some suggesting that he made a thousand percent on his investment. A further testament to the ability of the gurdies was the fact that many a miner left the Cariboo lamenting the fortune he'd spent on the dance girls. Such folks must have done plenty of dancing and drinking because that was all they'd get from the gurdies. They were upright, moral characters. In fact, more than a few eventually married miners and remained in the region when the Cariboo Gold Rush petered out.

Those miners intent on more than dance and drinks were not out of luck. It was easy enough to find someone to tickle a fancy in the Cariboo. Entrepreneurs in mining towns sought to take the edge off the physical work of the miners. One group adept at doing that was the prostitutes. If the *Cariboo Sentinel* was accurate in its description of the ladies of the evening, most miners would have needed a pretty sharp physical edge rounded off before they would seek out the services of these women.

"The prostitutes on the creek—nine in number—put on great airs. They dress in male attire and swagger through the saloons and miners camps with cigars or huge quids of tobacco in their mouths, cursing and swearing and look like anything but the angels in petticoats heaven intended they should be. Each has a revolver or bowie knife attached to her waist, and it is quite a common occurrence to see one or more women dressed in male attire playing poker in the saloons or drinking whisky at the bars. They are a degraded set, and all good men in the vicinity wish them hundreds of miles away."

Left unwritten was the fact that they needed the weapons because they were regular victims of assault and as often as not more likely to be short-changed as paid for their services. The "good men in the city" could only wish for the prostitutes' departure because of the more respectable parlor saloons (popular euphemism for brothels) and private houses operating in the towns along Williams Creek. While the tougher ladies congregated in more isolated areas (out of town or in Barkerville, in small, squalid structures behind Main Street near the creek), the establishments of the more feminine were out in the open and just as likely to be found next to a barber shop as attached to a saloon. While the names of the women who entertained the miners are mostly lost to history, the madams who ran the places were usually well known, if not well respected, members of the community. Most knew Mary Sheldon and Mrs. Nathan by name, but Fanny Bendixon was Barkerville's best known madam. Of French origin, she arrived in the Cariboo in 1865 following extended stays in San Francisco and Victoria. She immediately established the Parlor Saloon and, after that went belly-up in a few years, the Belle Union Station. A sharp businesswoman, she was never one to limit her businesses to satisfying a single physical appetite and her places always advertised the best in booze and cigars. Over the years she held interest in a handful of other establishments, so she prospered. In 1889 Judge Begbie, on a visit to the Cariboo, wrote of her, "Madame Bendixon is here in great form, indeed, enormous, vast, of undiscoverable girth, though she was always of goodly diameter." In her declining years, she brought out her grand niece, Leonie Fanny, to live with her as a companion. She was still living in Barkerville when she died in 1899, still a wealthy woman.

There were other women drawn to the Cariboo in the rush of '62, although they were few in number and they generally arrived in the years following the first great strike on Williams Creek. Some lonely miners weren't content with the occasional dalliance with a prostitute and preferred marriage. If a fellow was going to hook up, then he had to travel down to Victoria where there were more single women. The search was made easier after 1862 when the British Columbia Emigration Society began to bring in eligible young ladies. Many of these young women went on to form the nucleus of the stable society that eventually emerged in the larger towns of the Cariboo. The majority lived humble lives as miners' wives, but a handful opened or operated successful businesses, including boarding houses, hotels and saloons. And some observers noted that the prostitutes weren't the only women in the Cariboo who dressed in men's clothing. A handful of women possessed mining certificates and, though it was more common for them to have the licenses so as to allow their husbands a second claim, some took to mining.

But not all the miners were anxious to marry, as suggested by James Anderson. The last word on the matter of women up around Williams Creek is best left to him, as it appeared in his *Sawney's Letters*.

> *There are some women on this creek,*
> *Sae modest, and sae mild and meek!*
> *The deep red blush aye pents their cheek,*
> *They never swear but when they speak.*
> *Each ane's a mistress, too, ye'll find,*
> *To mak guid folks think that she's joined*
> *In honest wedlock unto one;*
> *She's yours or any other man's!*
> *But dinna fear, for me at least,*
> *I'll never mak mysel' a beast!*

There's little about mining that can be described as anything more than boring and exhausting. While the big strike brought relief to some, many miners took to passing long days by discussing the odd sights of the Cariboo. There was never a shortage of topics to chew on—the hanging judge who shouted like the devil and dressed like a scythe-bearing messenger of the dead, the women who dressed like men and spat tobacco with a snarl and a curse that flushed the stubble-covered cheeks of even hardened onlookers, or the crazy prospectors who mined where no one believed there was gold. But throughout the summer and fall of 1862 there were two new arrivals to the Cariboo who were greeted with open-mouthed disbelief from even the most cynical and experienced of miners. One was the Dromedary Express while the other was the Overlanders.

Entrepreneurs in the Cariboo, like those in every gold rush, were always on the lookout for a way to make a fast buck without actually digging for the yellow ore. Among the most profitable enterprises was supplying the miners but it came with expensive overhead, most of it consumed in freighting the goods into the interior. As the Cariboo Road was being built and travel by land became more accessible, the standard manner of delivering goods was by mule or mule train (which consisted of up to 50 animals). Mules carried from 250 to 400 pounds each; they needed plenty of watering breaks; and they traveled at an average rate of about 10 to 15 miles per day (depending on the size of the train). Most suppliers were content with the size of the loads and accepted the mules' slow pace and drinking needs as part of doing business. Frank Laumeister was not among their number.

In the spring of 1862 Laumeister and his associates imported 23 Bactrian camels from San Francisco. At $300 per camel the undertaking involved considerable expense, but Laumeister saw it as a sound investment. They were battle-tested animals, used by the United States army to pack supplies in the southern states. Reports suggested that they could carry 800 pounds each, could go for extended periods without water, and easily traveled 30 miles per day. It was all true and there was every reason to believe that the Dromedary Express (an inaccurate name, since the camels had two humps, not one), as it was quickly dubbed, would be a success. Yet Laumeister never realized the $50,000 of profit he envisioned pocketing in the first year of operation. While he knew plenty about grandiose money-making schemes, Laumeister knew precious little about camels and so the operation was a disaster.

The camels proved to be the most ornery beasts in British Columbia. They bit and kicked anything that came within assaulting distance. Staying beyond kicking range only meant that a fellow was more likely to be dripping great gobs of camel spit. Distance didn't seem to matter when it came to the stench of the beasts. Mules, horses, and even oxen bolted off the trails when they smelled the camels approaching. Most of the men who accompanied the fleeing animals soon scattered after them, clenching their noses as they ran. Added to the problems that they caused others was the camels' unsuitability for the Cariboo terrain—their hoofs could not bear the often rocky surfaces. Laumeister did his best to address the difficulties by employing deodorants for the stench and canvas and rawhide shoes to protect their hooves, but they were ineffective remedies. Facing threats of lawsuits from other pack train operators seeking compensation for their damaged animals and goods, Laumeister and his associates unceremoniously abandoned

Arizona entrepreneur and visionary, Frank Laumeister, surely thought that he had the solution to hauling freight into the Cariboo region when he imported a herd of Bactrian camels from San Francisco. Discovering that they could carry almost three times the weight an average mule could and that they could go without water for long periods, Laumeister was certain his investment would serve as his very own gold mine. He hadn't counted on their aroma (other animals balked at the overpowering smell), nor did he factor in their soft, spongy hooves being torn to shreds by the sharp rocks of the Cariboo Road. The problems proved insurmountable and the Dromedary Express was doomed. Two years later, Laumeister released them into the wild; the last one died in captivity in 1902.

the Dromedary Express in 1864. The camels were simply released into the wild. One was eventually killed for its meat even though no one would eat it, while another was shot for its hide and that wouldn't tan. The last of the camels died in 1905, but in the intervening 40 years they provided more than one heart-stopping encounter for unsuspecting travelers who stumbled upon them.

In the fall of 1862 there arrived in the Cariboo a group of fortune-seekers who had their own heart-stopping experiences. They were among many who left Britain's eastern North American colonies for a chance to try their luck in her western ones. Their exploits were to mark them as unique among the thousands determined to find gold in the Cariboo. Known as the Overlanders, these would-be miners set out for the Cariboo by traveling west across the continent. In total, they numbered less than 200 (smaller parties followed in the footsteps of the first large group). For the most part they were Canadians, though by journey's end there were British and Germans among their number. There were also two women and a handful of children. The Overlanders saw the transcontinental trek as an intelligent solution to the problems of distance, time and expense associated with the usual methods of travel to the Cariboo, but most folks who heard of their journey were more inclined to shake their heads and marvel at their craziness.

On the surface, traveling overland to the Cariboo appeared to be a sound decision. The distance from Halifax to Victoria, around the South American Horn, was a good 12,000 miles. Even a fast ship might take six months to cover it. Prospectors, concerned that the good claims might be longed staked while they were steaming along, could cut those figures well down if they went via Panama or Nicaragua (where a train would speed them from Atlantic to Pacific), but what was saved in time was expensive. Even if they could strike a favorable deal, the fare was still at least $500. In the face of such considerations, a trip across the continent was appealing. At only 3500 miles, most expected that a spring departure would be followed by a summer arrival. The shared costs of the expedition meant that per capita expenses would be minimized and, at about $100 a person, they were. Furthermore, the Overlanders had only to look south for both inspiration and example to guide them in their mission. During the California Gold Rush, many miners had traveled overland from eastern cities to the western gold fields.

The Overlanders were to discover that distance was a relative concept. The distance of 3500 miles might have been shorter, but traveling through often rugged and empty terrain was by no means easier. The example of their American counterparts was misleading. It's one thing to travel across a continent when the trails are well used and dotted with trading posts and the occasional city, depending on the route taken. It's another matter to travel through a vast unknown land, a fairly accurate description of the expanse of Rupert's Land leading up to the Rocky Mountains at mid-century. Such were the mysteries associated with the territory that it isn't much of a stretch to suggest that the Overlanders were their era's la Vérendrye or Fraser.

While most of the Overlanders officially began their western trek in early spring, from Montreal and Queenston, the challenge didn't really begin until Fort Garry (Winnipeg). The first leg of the journey was an easy one because the Overlanders used a combination of rail and steamboat through the American states, since there was no convenient access through British North America to the eastern edge of the Rupert's Land and the great northern prairies. Once in Fort Garry, they purchased supplies.

They had likely already bought personal effects, tents and mining equipment in the east, so the goods acquired at Fort Garry were mostly food. Flour and pemmican were especially important, but stocks of beans, codfish and dried apples filled out their stores. There were other expenses as well. Red River carts could be obtained for $8 each, horses for $40 a head, and oxen for $25 an animal. The Overlanders weren't so naive as to think that they could reach their destination unaided, so they also hired four guides.

On June 2 the Overlanders left Fort Garry. The expedition included 97 carts and 110 animals and, strung out along the prairie, it spanned a good half mile. From the first the group was determined that theirs would be an organized undertaking. To that end, they appointed a captain whose main responsibility was the general management of the train (camp discipline, how long to travel and at what speed, where to camp). Thomas McMicking of the Queenston party was the unanimous choice to fill the position. Assisting him was a committee composed of men who represented each of the smaller groups who together made up the expedition.

For the first month, travel was relatively easy. They made about two and a half miles an hour, a reasonable 25 miles a day. Nevertheless, it was a long day, starting at 2:30 AM and ending about 8 PM (depending on where they found water). For the most part, their only troubles were the merciless clouds of mosquitoes and the stagnant pools that served as their sources of water. When strained and well boiled, the soupy liquid didn't cause too many stomach ailments. Generally, the party was well rested because, by consensus, they had decided not to travel on Sundays.

During those first weeks, spirits were so high that musical instruments and vocal recitals were common events, but the journey took a sour turn after the Overlanders left Fort Pitt

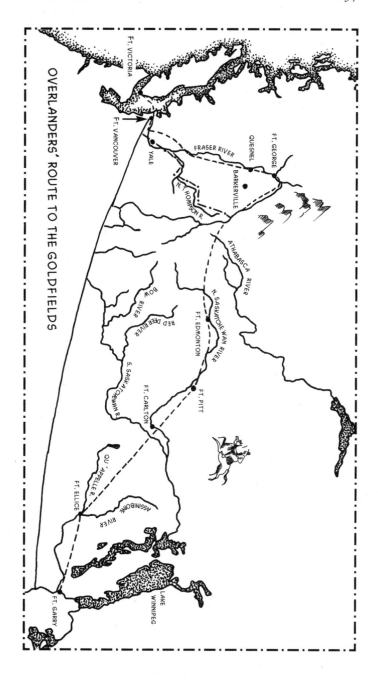

on the North Saskatchewan River in the second week of July. It foreshadowed the troubles to follow, beginning with eleven days of continuous rain that swelled all the streams, making passage difficult. Within three days they were forced to bring the party to a halt eight times in order to build bridges, some of them 100 feet long. When there was no timber, as so often happened on the prairies, they filled the creek with their carts and made a bridge of cart boxes on the top. Fortunately, these tasks were made easier by the presence of engineers in the expedition. The best that could be said for the unexpected work was that it gave the men a job on which to focus and thereby put a temporary end to the squabbles, ugly words and simmering discontent that weeks of close quarters had finally brought to the surface. By July 21 they were in Fort Edmonton.

When the Overlanders left the fort, they no longer had their Red River carts. Although they didn't know much about the mountains, they did know that the big-wheeled wagons would be of little value there. They were also 25 fewer in number because that many had decided to prospect on the North Saskatchewan. It wasn't long before others wished they had made the same decision. By early August, the elevation was getting higher, the nights colder and their food supplies were dwindling. By mid-August they were desperate and they took to shooting and eating their pack animals. Occasionally they turned to more dubious food sources, as related by McMicking. "We dined this day upon a dish so delicate and rare that it might have tempted the palate of Epicurus himself; so nice, indeed, was it, that I have little hesitation in naming it, lest we might be censured for living too luxuriously be the way. It was a roasted skunk.... After we had finished our repast, which all appeared to relish, we wondered that we had not discovered its good qualities sooner, and unanimously resolved, that his skunkship had been a slandered and much abused individual." When they finally reached Tête Jaune Cache on the Fraser River in late August, they encountered a friendly band of Shuswap who were willing to trade food for the Overlanders' non-perishables.

Here the party split into two groups. The smaller contingent (32 in total) took the horses to Fort Kamloops and the remainder were determined to ride the Fraser to Quesnel. While they strapped together rafts, carved dugouts and stitched together green hides fashioned around willow frames, the Kamloops-bound party departed. After a few weeks spent hacking out a trail through the dense forest without the certainty that they were actually headed in the right direction, they abandoned their mission and made for the Fraser. They slaughtered their animals, built some rafts and pushed off onto the river. Within a few days they encountered the Mouth of Hell, a narrow canyon through which the Fraser roared with merciless ferocity. They lost one of their rafts, much of their supplies and one man. Eventually they pulled their rafts ashore and resumed their pedestrian journey. Somehow, in mid-October they finally stumbled into Fort Kamloops. It was none too soon. Two days after their arrival, the woman among their party gave birth!

The Overlanders who set out for Quesnel didn't fare much better. The Fraser took its toll—two died. Rafts were swamped or splintered and supplies swept away. McMicking vividly described the all too common challenges of the Fraser, "Onward they [those on one of the vessels] sped like an arrow. They seemed to be rushing into the jaws of death. Before them on the right rose a rocky reef against which the furious flood was lashing itself into foam, threatening instant and unavoidable destruction, and on the other side a seething and eddying whirlpool was ready to engulf in its greedy vortex any mortal who might venture within. With

The lure of gold drove the legendary Overlanders to do what the North-West Mounted Police would not do for another 11 years: make the cross-country trek from the British North American colonies to the Rocky Mountains. In fact, many of the Overlanders went farther than the Mounties, continuing on past the Rocky Mountains after completing the arduous journey across the forbidding plains. Perhaps there is no better example of the effect gold fever had on people's minds. Above is a depiction of the final leg of the excursion for one group of Overlanders; several men struggle to make their way down the roaring rapids of the Fraser River. Two men died while trying to pilot these makeshift vessels towards the Quesnel River.

The Cariboo

fearful velocity they were hurried along directly towards the fatal rock. Their ruin seemed inevitable. It was a moment of painful suspense. Not a word was spoken except the necessary orders of the pilot.... Now was the critical moment. Everyone bent manfully to his oar. The raft shot closely past the rock, tearing away the stern rowlock, and glided safely into the eddy below. The agony was over. The gauntlet had been run and all survived." As the Overlanders were to discover, the Fraser had many such gauntlets and much of the journey was taken with heart in mouth.

On September 11, what was left of the party reached Quesnel. As was the case with the party that found Fort Kamloops, only about half of the Overlanders chose to remain and mine in the Cariboo. Tired and disheartened, the remainder headed for Victoria, evidently sharing the acerbic evaluation of the journey as summed up by one of the participants, "Our mining tools were the only articles that we found to be unnecessary."

In terms of both participation and production the Cariboo Gold Rush reached its peak in 1863. While there were never as many miners in the Cariboo as had prospected along the Fraser in the late 1850s, they still numbered in the thousands. Undoubtedly there would have been many fewer had it not been for the Cariboo Road. Even though it wasn't completed in 1863, its very construction drew many to the interior who might have thought otherwise given the hostile geography. Most of them ended up around Williams Creek and of those who did, few were not involved in mining "deep leads" (shaft-sinking and tunneling). And why not? The returns from the physical demands of digging holes were staggering. During the boom year of 1863, miners took nearly 250,000 ounces of gold from the region.

Given the bountiful deposits, it's hardly surprising that there were plenty who became incredibly rich. Some were just plain lucky, like Billy Barker. Other times fortune came by way of a sharp eye. H.F. Davis was one of the ill-starred prospectors who arrived in Williams Creek after all the good ground had been staked. As he sat and cursed his luck, his eyes roamed over the claims. One looked like it was a little larger than the regulation 100 feet. Following up on his suspicion, he discovered that it was 112 feet. He immediately claimed the excess footage and soon mined $12,000 worth of gold from it. His success won him the nickname of "Twelve Foot" Davis.

Of course, the stories of all gold rushes are incomplete without the tales of tragedy. The sad ending of the handful of Overlanders provides only part of the Cariboo's chapter on the subject. There were many other bleak stories. John Angus "Cariboo" Cameron's story fills a few more pages. With his wife and 14-month-old baby, he headed for the Cariboo in late 1861. The long ocean voyage proved too much for his daughter and she died upon their arrival in Victoria. He had to borrow money to get a stake together, and by fall he and his wife were in the inland gold fields. While he prospected unsuccessfully, his wife came down with typhoid fever and died. It wasn't long after that he and his partners struck pay dirt; they discovered the Cariboo's second richest claim. He left the Cariboo with $350,000, but no family.

By the mid 1860s, most miners had moved on from the Cariboo. The American Civil War was raging and plenty signed on for the steady wage. Others moved on to the new rush in southern British Columbia around Wild Horse Creek, but that one never came close to the bonanza of either

The Cariboo

the Fraser or the Cariboo. While the Cariboo boom times lasted only a few years, the strike was to echo though the years that followed. The gold rush and Governor Douglas' efforts to deal with it did much to promote and solidify what had been an undeveloped and pretty much insignificant colony. When the Dominion of Canada was created in 1867, British Columbia (by then a single colony) was courted in an effort to establish a nation that stretched from sea to sea. It took a few years for B.C. to accept the proposal. By then, Douglas had been replaced as Governor (the Cariboo Road had cost $1.25 million, all of it unauthorized spending), but folks could be forgiven if they considered him one of the Fathers of Confederation.

The Black Hills
(1876)

GENERAL GEORGE ARMSTRONG CUSTER retired to his tent at the main camp of the Black Hills Expedition in Prospect Valley, Dakota. Once inside, he sat before a table and began to write out a report to Lieutenant General Philip H. Sheridan, his division commander. Custer was by nature given to action rather than thoughtfulness but he uncharacteristically devoted considerable reflection to this report. More specifically, he thought long and hard about the best way to tell Sheridan of the gold deposits discovered by the miners associated with the expedition.

During the first weeks of the expedition, the miners who traveled with Custer's command, Horatio Walsh and William McKay, had been disappointed to discover that the little gold they found in the creeks of the Black Hills wasn't much more than a miner might wash out of any western stream. Then they stumbled upon French Creek, just south of present-day Custer, South Dakota. Their frustration turned to enthusiasm when they discovered excellent colors, a few flakes of gold, in the loose soil on

George Armstrong Custer (1839–1876) graduated—barely—from the prestigious military academy at West Point. He distinguished himself during the Civil War and in 1866, he was posted on the Plains, certain that in the subjugation of the Natives he would find the glory he so desperately sought. When General Philip Sheridan was made commander of the Department of the Missouri, he was charged with bringing the Plains Natives under control. Sheridan gave Custer command of the Seventh Cavalry. In 1873, Custer led an expedition to explore the Black Hills. Finding the land well suited for agriculture, he also found gold. The American army's questionable incursion into the Dakota Territory and the subsequent unlawful rush of miners into the region raised the ire of the Sioux and resulted in many deaths, both Native and white. But none was more famous than Custer's in his last stand at the Battle of the Little Bighorn.

the stream's shore. They estimated that the dirt would pan out at about $.10. In such amounts, a determined miner could bring in $50 to $75 per day. Walsh and McKay were also quick to point out that the Black Hills were marbled with the veins of quartz that were a telltale sign of gold deposits. With time an issue, they didn't do a thorough investigation of the veins because that demanded the labor intensive process of post-holing.

Custer stared at the gold sample that Walsh and McKay had given him. Most men would look at gold and see a fortune. All that was visible to Custer was trouble. The last thing he, or the United States army, desired was a mass stampede by miners into the Black Hills. The region was Sioux territory, and Custer knew from experience that they were not likely to welcome white interlopers. Heaving a great sigh, he put pen to paper.

"Gold has been found at several places; and it is believed, by those who are giving their attention to this subject, that it will be found in paying quantities. I have upon my table 40 or 50 small particles of pure gold, in size averaging a small pin head, and most of it obtained today from one panful of earth. As we have never remained in camp for longer than one day, it will be readily understood that there is no opportunity to make a satisfactory examination in regard to deposits of valuable minerals. Veins of lead and strong indications of the existence of silver have been found. Until further examination is made regarding the richness of gold, no opinion should be formed."

Custer added some details about the terrain and the expedition's mapping of it and then read the report. Satisfied that any optimism it suggested was tempered with an appropriate degree of prudence, he signed it and had it sealed for delivery with the next package of dispatches.

> **THERE'S GOLD IN THEM THERE HILLS!**
>
> **CUSTER REPORTS GOLD IN THE BLACK HILLS**
>
> In a report received from the expedition led by General George Armstrong Custer, he says that the miners associated with his expedition have discovered the presence of gold in French Creek, in the Dakota Territory.

The general's care in preparing his dispatch went for naught because there were those on his expedition whose business it was to sensationalize news rather than simply convey it. Reporters constituted a sizeable proportion of the non-military contingent that accompanied his command. They represented newspapers located in several American cities, including New York, Chicago and Bismarck. And they weren't interested in what the army declared were the purposes of the expedition—to conduct reconnaissance and to determine the most suitable location to build a fort. No. The glitter of gold was what sold newspapers.

In August 1874, America was introduced to the discoveries. Great booming headlines in the *Yankton Press and Dakotaian* proclaimed "Struck It At Last! Rich Mines of Gold and Silver Reported Found By Custer." The headlines optimistically continued, "Prepare For Lively Times! Gold Expected to Fall 10 Per Cent—Spades and Picks Rising — The National Debt to be Paid when Custer Returns." While some reporters cautioned the prospective miner to be wary and wait for a more extensive study of the mineral deposits in the region, the more common attitude was that reported in the *St. Paul Daily Press*, "Custer's Expedition.

Where Gold and Silver Lie Scattered Around, and which is Already Making Plethoric Knapsacks of the Boys." The report declared, "From indications this is the Eldorado [sic] of America."

Pleas for restraint from the United States army could not turn men's eyes from the sparkling dreams of El Dorado. The rush was on. And Custer, for one, would regret the day he set foot in the Black Hills.

∽∾

The Sioux called the Black Hills *Paha Sapa* (hills that are black), but the terse description suggested nothing of the high regard in which they held them. They considered those hills to be sacred ground. It was where their gods lived and, in their mythology, it was the place of the tribe's creation. Children sat around evening fires and listened to the stories of *Wakan Tanka*, the Great Spirit, and of how, in a moment of anger against the people, he commanded *Unktehi*, the water monster, to flood the land. Their ancestors drowned in the chasms at the summits of the *Paha Sapa*. A new nation sprang forth from the union of a surviving maiden and *Wanblee Galeshka*, the great eagle. The offspring of the couple were known as the Sioux, the brave buffalo hunters of the plains. For as long as the elders could recall, the Sioux had continued to live around *Paha Sapa*. Periodically their journeys took them far from the shadow of the mystical hills, but they always returned.

Paha Sapa had more than spiritual significance for the Sioux. The mountains bestowed abundant resources. Forests blanketed the hillsides providing both firewood and suitable poles for tipis. Game found shelter and security beneath the branches of the trees and multiplied. Rivers flowed through

The Sioux had long called them *Paha Sapa*, the mountains cradled by the Belle Fourche and Cheyenne rivers. European explorers thought the name apt and used the English translation, the Black Hills. For the Sioux, the group of wooded hills rising up from the Great Plains was sacred ground, providing them with material and spiritual sustenance. It was the mythical birthplace of their people and a sure place to find game when times were lean and comfort in times of uncertainty. Yet, with the news of the mineral wealth buried in the hills, prospectors saw only one thing in the lush hills: gold. The Black Hills were the perfect symbol of the struggle between the indigenous peoples and the American invaders. Both groups were looking at the same landscape, but they saw radically different things there. Before long, they would go to war over these differences.

the valleys fed by clear mountain streams. The great Sioux chief Sitting Bull referred to *Paha Sapa* as a food pack. It was a curious statement but every Sioux knew what he meant—when they were in need of something, material or spiritual, they could just go to the mountains and get it. For the Sioux, *Paha Sapa* was worth protecting.

When the Sioux set their mind to something, they were determined and they used whatever resources were at their disposal to achieve the objective. Their greatest resource was their courage. The Sioux were known across the depth and breadth of the plains for their bravery. Boys were raised on stories of war and longed for the day they could count coup (strike an enemy with a coup stick) and become warriors. Men liberally covered themselves in war paint and enthusiastically ran the daring line (the open space between the two lines of opposing forces) intent on gaining the respect of fellow braves. Sioux culture was rich and complicated, but at its heart was a undaunting sense of bravery that made them fierce opponents, as Americans who wanted *Paha Sapa* for themselves were soon to discover.

It was well into the 18th century before the white man knew anything of *Paha Sapa*. The la Vérendrye brothers, Louis-Joseph and Francois, were likely the first Europeans to see the Black Hills. Louis-Joseph la Vérendrye led the first European exploration into the region in the 1740s. They may have been puzzled by what indeed appeared to be black hills on the horizon. From a distance the dark green deciduous trees that mantled the sloping mountains seemed black enough. The la Vérendryes didn't get close enough to investigate the puzzle. They were only able to circle the mountainous region cradled by the branching Belle Fourche and South Fork Rivers because their Native guides refused to take them into the territory. With the Natives' stubborn refusal was born the mystery that swirled around white explorers' accounts of the Black Hills for more than a century. In the decades that followed the near encounter of the la Vérendrye's, a handful of other explorers saw or were told of the hills. But it wasn't until Jedediah Smith and his band of fur traders penetrated the southern reaches of the range in 1823 that a white man actually set a foot in the sacred territory. While there, Smith almost lost his scalp…to a grizzly bear.

After Smith, most of the European exploration was left to prospectors. Ezra Kind was among the first of those. With six companions, he traveled north from Fort Laramie on the Platte River in 1833. They found gold alright, but they also found that they weren't able to spend it. The seven perished, the only eerie record of their demise left scratched by Kind on a rock, "Got all of the gold we could carry our ponies got by the Indians I have lost my gun and nothing to eat and Indians hunting me." It's likely that the Sioux killed the Kind party because they were where they should not have been and not because they were searching for gold. However, the Sioux soon learned to be wary of interlopers prospecting for the yellow ore. In 1848, the respected Jesuit missionary Father De Smet was shown samples of gold collected by the Sioux. He advised them to guard their secret well lest the greed of the white man destroy them. The Sioux took his words to heart and pledged to kill any of their own who showed the gold of *Paha Sapa* to the white man. Any who came looking were to suffer the same fate. In the early 1850s, when another party of 30 men prospected for gold in the hills, only eight returned with reports of gold. The remainder were never heard from again, although two of their skeletons were subsequently found.

In the years that followed there were more expeditions. Sometimes they were carried out by wealthy amateurs, like

Sir Saint George Gore in 1855. The reports from Gore echoed the few facts and many rumors of earlier expeditions—there was gold in the Black Hills. To determine the accuracy of the assertions, the first scientific expedition was dispatched in July 1857. Led by Lieutenant-Governor Kemble Warren and supported by a military escort of 17, the expedition was forced to turn back at Inyan Kara Mountain just west of the Black Hills. The Sioux were camped there and insisted that he go no farther. The army organized a more determined military operation in 1865. The Powder River Campaign was coordinated into three successive expeditions. Concerns about surprise attacks by the Sioux prevented any of them from venturing into the heart of the Black Hills, but each expedition explored the fringes of the region. Each also reported discovering gold deposits. By then, the presence of gold was no longer in doubt. Some Sioux, apparently having abandoned their pledge, were trading it at Fort Laramie. The only remaining question was the quantity in which it might be found.

Despite the news of gold in the Black Hills, no one rushed to mine the territory. It was isolated, without easy access and not a preferred choice when there were other, more profitable mining fields to exploit—diamonds and gold in South Africa, silver in Nevada and gold in the Cariboo. The times were also against a rush because many of the men who might have turned to prospecting were fighting in the American Civil War. Undoubtedly, however, the strongest deterrent was the Sioux, who had vaulted onto the American stage as the most terrifying of western Natives.

Throughout the late 1850s and 1860s, as the trickle of westerly-bound settlers exploded into a torrent, the Sioux began a campaign of resistance that turned the Midwest red with blood. It started in Minnesota with the Santee Sioux. In the early 1850s they signed a treaty with the American government. In return for food, equipment and money for trade goods, they agreed to confine themselves to a narrow strip of land above the upper Minnesota River. A sorry pattern soon emerged. The annuities were regularly late and when the money finally arrived, unscrupulous traders and Indian agents callously charged outrageously high prices, bilking the Natives of what was rightfully theirs. The Sioux responded by killing 40 settlers. It was a precursor of what was to come.

In 1862 the new president, Abraham Lincoln, signed the Homestead Act authorizing grants of public lands to homesteaders who tilled it for five years. Thousands packed up and set out for the Overland Trail through the northern territories, the homeland of the Santee Sioux. Rolling wagons were driven across the countryside without the slightest concern that they were trespassing. Exasperated, the Santee took to protecting their territory, and violence became the norm from Minnesota west to the Dakotas and south into Iowa. Eventually some 800 whites lay dead, only one hundred of them soldiers. Hundreds more were taken prisoner. The United States army replied with force. Led by Henry H. Sibley, the Americans defeated the Santee chief Little Crow at Wood Lake, Minnesota. More than 400 Santee were captured and, of these, a commission sentenced 300 to hang. Frontier justice gave way to the more reasoned approach of President Lincoln, who personally reviewed the reports dealing with the condemned Santee. He revoked the death sentence of all but 38.

To the south in the Powder River country, the Oglala Sioux were just as occupied with the newcomers. For many years, gold-seekers who sought access to the mines in Montana had used the Missouri River as a point of entry. It was a long, exhausting route. In 1864, John Bozeman established a shortcut that linked the Platte River with the mines. The Bozeman Trail quickly became the most popular route of access and, in 1866, the army built three posts along the trail—Forts Reno, Phil Kearny and C.F. Smith. The trail and the posts were a little too close for the Oglala and they took to arms. In December 1866, led by Red Cloud and Crazy Horse, they wiped out Captain William Fetterman's unit of 80 men.

Other Sioux bands joined in the resistance and, while it couldn't be said to have been an organized response, all the bands shared a similar motivation—none wanted to see American settlers take their land. Anxious to avoid a protracted, costly and violent Native war, and perhaps inclined to take a moral high road, the American government sat down and negotiated with the Sioux. The result was the Fort Laramie Treaty (the Treaty of 1868) that proclaimed peace between the Sioux and the American government. Central to the treaty was the closing of the Bozeman Trail and the creation of the Great Sioux Reservation, which included all of what would become the state of South Dakota west of the Missouri River. In the southwestern corner of the reservation were the Black Hills.

The terms of the treaty stipulated that the Great Sioux Reservation was to be "set apart for the absolute and undisturbed use and occupation of the Indians named herein...." The American government went on to agree that "no persons except those herein designated and authorized so to do, and except such officers, agents and employes [sic] of the government as may be authorized to enter upon Indian reservations in discharge of duties enjoined by law, shall ever be permitted to pass over, settle upon, or reside in the territory described...." It appeared that the Sioux had won protection for their food pack, the sacred *Paha Sapa*.

The Black Hills

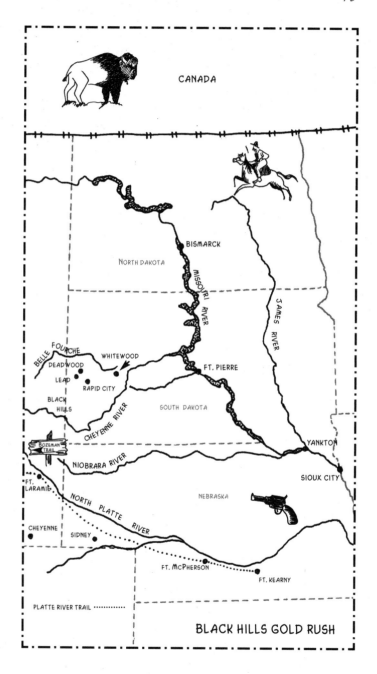

Black Hills Gold Rush

For a while the American government kept its promise. Efforts by Dakotan businesses to exploit the resources of the Black Hills were firmly squashed by officials in Washington, but pressure continued to mount. When hostile Sioux (those bands who hadn't signed the Laramie treaty) continued to attack settlers in the region, the army used it as an excuse to further explore the Black Hills with the intent of establishing a post there. Justifying the reconnaissance by asserting that the expedition was comprised of government officials legally authorized to enter the reservation, General William T. Sherman, the commanding general of the army, approved the expedition. George Custer was authorized to lead ten companies of the Seventh Cavalry into the Black Hills. When the expedition set out on July 2, 1874, Custer had under his command 951 soldiers and teamsters, and a good collection of civilian aides, including reporters, miners, scientists and a photographer. They weren't sure what they'd find, but they made certain that they were prepared for trouble.

The trouble that Custer anticipated donned war paint and rode bareback. As his expedition wound its way into the Black Hills, he was surprised to discover that there were few Sioux there. He shouldn't have been. As Sitting Bull had noted, the Sioux only used the place in times of need; for the rest of the time the spirits were left to live in peaceful solitude. Custer found trouble nonetheless. It came in the tiny sparkling pieces of metal found by the miners who accompanied the expedition.

When Custer wrote his words of caution on the matter of gold deposits, he may as well have been spitting into the

The Black Hills

Although the United States Army was ostensibly committed to a policy of peace with the Natives, their actions on the Plains through much of the last half of the 19th century were anything but peaceful. Even after they signed the Fort Laramie Treaty, the army continued to make inroads into Sioux territory. The photograph above was taken of Custer's reconnaissance into the region. Officially, Custer's mission was to survey the landscape in order to determine the ideal location for another military fort, but the Black Hills Expedition was also charged with confirming reports of gold deposits. Everyone knew what would follow when the journalists in the expedition reported the prodigious amounts of gold found in the hills. The announcement set in motion the final major engagement between the American Army and the Sioux, culminating in the well-known Battle of the Little Bighorn.

wind. Even forbidding tales from some of the journalists accompanying the expedition warning that there was evidence to suggest that prospecting in the Black Hills would result in an untimely death rather than a gold strike, had precious little effect. As Samuel Barrows of the *New York Tribune* ominously wrote, "Those who seek the Hills only for gold must be prepared to take their chances.... The Black Hills... are not without ready-made monuments for the martyrs who may perish in their peaks." Such advice was lost in the enthusiastic stories more commonly written by journalists. Newspapers and magazines from across the nation were more inclined to reprint the intoxicating observations of William Eleroy Curtis of the *Chicago Inter-Ocean*, who declared that the Black Hills was the land of promise, "From the grassroots down it was 'pay dirt'...." And, when Charley Reynolds left the Black Hills Expedition to deliver Custer's dispatches to army officials and confirmed to reporters in Sioux City that he had actually seen gold washed from the dirt that filled the creek beds, well, the craziness began.

Sioux City boosters and businessmen were the first to try and exploit the findings. There were other cities closer to the gold country, including Cheyenne, Bismarck and Yankton, but none were quicker off the mark. Sioux City was southeast of the Black Hills and, located on the Missouri River, it had easy, if distant, access to the region. Businessmen had long been anxious to make a profit from the Black Hills, but efforts and plans were regularly turned aside by the army that stood by the terms of the '68 treaty for a few years at least. News of gold, however, shifted commercial wheels into full gear. Thomas Russell, the presiding officer of the Society of Black Hills Pioneers, with the able assistance of Charles Collins, the editor of the *Sioux City Times*, set up an office in Chicago from where the gold find was promoted with circulars and letters. Bankrolled by Sioux City capital, Russell advertised that $50 would get fortune seekers to the Black Hills, via Sioux City, of course. Within a few weeks, General Sheridan (commander of the Military Division of the Missouri), whose office happened to be in Chicago, issued an order prohibiting the proposed invasion of the Sioux reservation.

The promoters issued their own public announcement to the press; the excursion was off. Secretly, however, private letters followed the declaration, "If you can raise $300, can handle a rifle, and mean business, be at Sioux City on or about the middle of September." Local merchants were busy making preparations for the expected arrival of the adventurers. A few tents were set up and the newcomers were advised that $100 would outfit them properly for the journey. They would get pick, shovel, gold pan, cooking utensils, blankets, and, more ominously rifle, revolver, and sufficient ammunition. By September, the population of Sioux City increased by 100 prospectors and local wallets fattened proportionately.

Despite the "secrecy," General Sheridan knew about the planned excursion and he was well aware of the problems it was likely to bring. If men were foolish enough to risk their lives, so be it; but he also knew that the Sioux weren't likely to stop fighting when the miners were dead. If they went on the warpath, the finer points of picks, shovels and western-bound wagons would be lost, and any whites they stumbled upon would be seen as fair game. Intent on keeping the reservation free of white interlopers, Sheridan issued orders to his commanders in the region to stop the expedition—burn the wagon trains, destroy the outfits, arrest the leaders, but stop the expedition. Army patrols set out to keep the routes to the Black Hills clear. Sheridan's strategy had its effect. By October, when the expedition finally set out, there were fewer than 30, including a woman and a boy. They were a determined party. That both the Sioux and winter awaited their arrival was no deterrent. Visions of gold had a way of clouding people's judgement.

While Russell accompanied the expedition, John Gordon was paid handsomely ($1000) to be its guide and captain, so it quickly came to be known as the Gordon party. Counted among the group were greenhorns and veteran frontiersmen, although it seems that none had any prospecting experience. They were cunning enough, though. The party had six canvas-covered wagons and painted each of them with the declaration "O'Neill's Colony," a settlement in Nebraska that the prospectors hoped curious eyes would assume to be their destination. Once they arrived in Sioux country, Gordon gave further orders designed to avoid the army. Once dusk fell there were to be neither fires nor loud noises. A mule with the nasty habit of braying at the moon was muzzled. Gordon also posted a rotating guard, likely more out of fear of raiding Sioux war parties than to look out for army patrols.

The journey was not uneventful, but as it turned out, the members of the party had more to fear from themselves than either the Sioux or the army. One fellow got so fed up with Gordon's dictatorial ways that he unloaded a shell full of lead at him. He missed and was lucky not to get a knife in the belly for his efforts. Here was proof enough that miners were an independent and violent lot.

The Gordon party first sighted the Black Hills on November 30. A few weeks later they were in the heart of the range, following Custer's trail. Just before Christmas they reached French Creek and the camp where Custer had found gold. They had already experienced an early winter blizzard and knew that prospecting would have to wait until adequate shelter was constructed. Judging by the stockade they built—80 feet square with walls of 13-foot logs set 3 feet deep and protective bastions at the corners allowing gunfire along the walls—it wasn't just the elements that concerned them. Inside the structure they built six cabins, dug a well and loaded up on firewood. The Gordon Stockade was completed by mid-January.

By then it was difficult to do much prospecting. The ground was frozen and any digging was nearly impossible. The streams were icing over and even the toughest could only handle a few hours of continuous panning, usually around midday when it was warmest. Precious little color showed, but it didn't dampen the enthusiasm of the prospectors. Though they found only about $40 worth of gold in all of January, they figured that men working full-time with rockers could easily take $10 a day. For the most part, they spent the weeks planning the town sites they were sure would swell up around their fortress. Nothing came of Henry City, but Custer, some three miles away, did become a reality.

With little to keep them busy and the next couple of months offering more of the same, in February John Gordon and one of his party decided to return to Sioux City. In March others fanned out to Fort Laramie and Cheyenne. They went for supplies and, more importantly, settlers. The colorful letters that filled their pockets surely kindled the fires of adventure in more than a few hearts. In a characteristic missive, R.R. Whitney cautioned that, while he didn't want to create too much excitement, his best advice to young men was to head for the Black Hills. He didn't mind giving people the lowdown on the region; he was set on bringing $150,000 out of the Black Hills and, despite the cold weather and the duties of stockade life, he was sure he'd reach his goal. Printed in newspapers across the land, news of Black Hills gold soon became common knowledge.

The presence of the Gordon party made General Sheridan hot under the collar. Despite recommendations from his commanders in the field that winter was not the best season to be sending troops after the trespassers, Sheridan was insistent. He wanted them removed. Through the winter, three details were dispatched and none of them enjoyed success. Two parties wandered through the hills and almost froze in the cold and blizzards. A third found the miners but let them be. Given their orders, it was an odd choice but it reflected the questionable attitude of the army (at least of those in the field) about enforcing treaty regulations. For the most part, neither soldiers nor miners gave enforcement of the regulations more than lip service.

In the spring there was a more determined attempt. An army detachment co-opted two of the Gordon party who had returned to Fort Laramie and had the pair guide them to the stockade. In early April, they arrived at the miners' base. The residents were given two days to pack up their gear and leave. No wagons were provided and they were forced to abandon most of the mining equipment, which was subsequently stolen by other miners and Natives. Such was the nature of prospecting. It was risky business and participants either made their fortunes or lost darn near everything. Most fell into the second category. Upon reaching Fort Laramie, the army simply released the miners.

Though it failed, the Gordon party was the impetus for the Black Hills Gold Rush. News of the Gordon party and the tales of gold that they spread kept the army busy over the next few months. Miners began to trickle into the region. Those who traveled in small groups of two or three managed to slip through the army's net with little trouble. Others, like Gordon's second expedition of nearly 180 people, were caught. When he refused to turn back, the army burned his wagons. Pressure from high-ranking army officials was clearly mounting. Perhaps sensing that the efforts of the army were going to make prospecting difficult, promoters turned to new ways to fill their pockets. One scam had would-be miners send in $5 apiece to reserve their place in the next expedition. When $3000 was collected, the promoters disappeared. While that wasn't a problem that the army had to deal with, they would soon enough face their biggest challenge. The Sioux had decided that they'd had enough of the army's lax enforcement of treaty regulations and would take matters into their own hands.

※

As the months of 1875 slipped by, more and more prospectors arrived at the handful of cities that formed a rough arc to the east and south of the Black Hills. Sioux City had gotten off to a quick start, but Yankton and Bismarck (both also on the Missouri River with the latter the western terminus of the Northern Pacific) and Sidney and Cheyenne (both on the Union Pacific route) were quick to join the party. For the most part, the wide-eyed newcomers they greeted were a desperate lot. Many had felt the sting of the nationwide depression of the early 1870s and they were anxious to start shaking a pan to shake off the economic doldrums. By late summer, reports of new finds were increasingly common. Names like Deadwood Gulch, Nigger Hill and Whitewood soon made eyes glaze over and hearts beat wildly. For many, relief could only be found in swinging a pick on their own claims in the Black Hills.

The army discovered that there were simply too many prospectors to allow for an effective patrol of the region.

Determined miners slipped into the hills with considerable ease and frustrating regularity. To stem the tide, General Crook, Sheridan's immediate subordinate in the region, issued a proclamation ordering miners to abandon the Black Hills. With a reputation as one of the most able Native fighters of his time, Crook was indeed an odd choice to defend Sioux interests. Perhaps then, it's not surprising that he did it half-heartedly. The proclamation directed miners to leave until new treaty arrangements with the Sioux were formalized. But they wouldn't leave empty-handed. Crook's proclamation also proposed a meeting of the miners where resolutions would be drafted securing for each his stake. Miners could reclaim their stakes, not if the Sioux were removed, but as Crook put it, "when this Country shall have been opened." Although there may have been some bleeding hearts in Washington who were opposed to riding roughshod over the Sioux, the attitude of the United States army was one of inevitability—come hell or high water, the Sioux would be driven from the Black Hills. Some miners agreed to the terms but most knew that the flaw in Crook's proposal was that there was no way of knowing when the Sioux would be gone. So the doubtful stayed and more kept coming. By late 1875 there were more than 500 prospecting in the hills.

Their presence made an unequivocal statement: the Fort Laramie Treaty of '68 was falling apart. From Sioux perspective, Custer's Black Hills Expedition was evidence enough that the white men would not be held to its terms. The steady flow of miners that soon followed the expedition only confirmed their belief. In their own defense, army brass claimed that the hands of the Sioux were hardly clean. They readily pointed out that in the half dozen years following 1868 the Sioux had killed more than 200 whites in territories as distant as the Missouri,

Yellowstone and Niobrara (Nebraska) Rivers. The Sioux could complain all they wanted but, as far as the army was concerned, they were a bunch of uncivilized hypocrites. It was hardly a just accusation since most of the attacks, if not all, were waged by Sioux who had not signed the treaty of '68 or who believed they were not subject to its terms (the Sioux rarely operated as a single entity with one spokesman as nation states did). The distinction was lost on the army. Nevertheless, there was no appetite for a large scale Native war so government officials ordered more negotiations.

In the fall of 1875, the Allison Commission met with Sioux chiefs and offered to buy the Black Hills. To ensure an atmosphere conducive to bartering, Allison pointed out that, despite the efforts of the army, more prospectors were inevitable. That could only mean more trouble for the Sioux. The Sioux, however, already knew something about trouble so threats of its increase didn't erode their conviction. But the presence of the miners did give them pause for thought. While the Sioux weren't interested in digging for gold, they knew that the white men were strangely anxious to do just that. The Sioux, like most Native peoples, were nothing if not good bargainers. They demanded $70,000,000 for the Black Hills. Outraged, the commissioners countered with an offer of $6,000,000 so, not surprisingly, left without a deal. Interestingly, by 1880, $14,000,000 worth of gold had been extracted from the Black Hills. Over the next half century, that number would increase to more than $400,000,000.

With the failure of negotiations and no foreseeable end to the violence, matters finally came to a head in 1876. Early in the year the government ordered the Sioux to return to their reservations. Failing that, they would be considered hostile and full scale military operations would be set in motion to force them back. The government was serious. In March, the army attacked a Sioux and Cheyenne winter camp. While few were killed, the village was razed and the Natives were left to face the rest of winter without supplies or shelter. The Battle of Powder River, as it became known, was a surprise attack on peaceful Natives. It enraged many Sioux. Young braves longed for the word from their chiefs that would allow them to fight back, but it was not forthcoming. Sioux leaders were determined to try to keep the peace; they would abide by their long standing custom and only raise their arms in self-defense.

However, the new policy of the American government and the restlessness of the young braves pressured the Sioux to reconsider their position. Sitting Bull, respected Hunkpapa Sioux war chief and medicine man, called for a war council. During the spring of 1876, Sioux and Cheyenne rode north to his encampment near the Yellowstone River. Among them were Crazy Horse and his band of southern Oglala Sioux as well as Spotted Eagle and his Sans Arc. When the council met, Sitting Bull was proclaimed the war chief for all his people. It was an unusual action but the times demanded a radical approach. One of Sitting Bull's first declarations was to call for a Sun Dance. During the ritual he would seek a vision from *Wakan Tanka*, the Great Spirit, that might provide guidance. It would be the most important dance in which Sitting Bull had ever participated. He prepared well, sweating in the lodge, smoking his sacred pipe, meditating. Finally, he was ready to offer his red blanket to *Wakan Tanka*. He had both of his arms pierced 50 times each until they ran red with blood. Then he danced around the great tree that served as the sacred center pole for the ritual. He danced through the night, finally collapsing the next morning.

For his efforts Sitting Bull was granted a vision. He heard a voice commanding him to look below the sun. As he did he saw many soldiers falling to the ground in the Sioux camp, their heads below their feet. Under them were some Natives, also upside down. Again the voice spoke to him. "These Long Knives do not have ears. They will die, but do not take their belongings."

And so it happened. First, the Sioux and their allies met the army at the battle of Rosebud Creek in mid-June. Led by Three Stars Crook, as the Natives called the general because of the number of stars on his uniform, the much larger army contingent was forced into retreat. Less than two weeks later the Seventh Cavalry, under the command of Custer, was wiped out at the Battle of the Little Bighorn. Custer's demise was as much a consequence of his own stubbornness and dreams of grandeur as it was of Sioux strategy, but it made no difference to the result. Following the battle, Sitting Bull fled north, eventually going to Canada. Crazy Horse returned to his homeland in the south. On the way, he took to attacking miners in the Black Hills. Through mid-summer, he was pursued by a vengeful Crook and, by September, the army had driven him from the region.

Crook remained concerned, however, that the miners might suffer continued attacks, so he posted his men in the Black Hills where they remained until late fall. A military post was also established near Bear Butte. Camp Sturgis, as it was named, ensured that the northern reaches of the Black Hills would be safe from attack from Sitting Bull. In addition, Crook suggested to the miners that they might want to establish their own militia but, as it turned out, the volunteer force was not necessary. In September, Washington dispatched a new commission to the Sioux territory, charged with obtaining the Black Hills. The Natives

Sitting Bull was North America's best-known and most-feared native warrior. For many years, he and his band of Hunkpapa Sioux harassed and attacked army camps and settlers to make it clear that whites were not welcome in Sioux homeland. In June 1876, Sitting Bull won his historic victory at the Battle of the Little Bighorn in Montana. Ten years later, Sitting Bull and his beleaguered band traveled to Canada assuming the Canadian government would protect them. But the Canadians saw Sitting Bull's presence as both a financial and political liability, so refused to provide his desperate band amnesty. After five years, the starving Sioux had no choice but to return to America. Sitting Bull was among the last to surrender. In December 1890, government officials ordered him arrested, because they were fearful of his influence over his people during the Ghost Dance. Sitting Bull was killed in the ensuing gunfight.

were promised that the delivery of the supplies and annuities guaranteed by the Fort Laramie Treaty of '68, which had been cut off during the hostilities, would be resumed. Predictably, the ultimatum was enough. The great buffalo herds on which the Sioux had once depended for their survival were a thing of the past and they were increasingly reliant on government supplies. They held their noses and signed the treaty. By February of 1877, the Black Hills were American territory and there was no longer any impediment to the rush.

The fact was that determined prospectors had never considered the Sioux to be much of an impediment. They came in such numbers and with such speed that they simply couldn't have been too worried about the Sioux, even though General William T. Sherman, the commander of the U.S. army, had characterized them as the most warlike nation of Natives on the continent. By May 1876, the population in the Black Hills was estimated to be over 15,000, although that was probably on the high side. The majority were to be found in more urban settings like Deadwood, Central City and Central Hills, but thousands roamed the valleys looking for the next big strike. Some found it.

In the spring, brothers Moses and Fred Manuel made their claim in a place called Lead. Before long they found a 200-pound nugget of gold quartz. They named their claim the Homestake and sold it for $100,000 the next year. The investment was a good one because over the next 85 years it produced more than $700,000,000 worth of gold. Nigger Hill was so called because two Black miners took $50,000 out of a gravel bank using only an old washtub and a few feet of plank flume to redirect the flowing water. In Rockerville Gulch, a wheelbarrow of dirt was said to produce $100.

This is a view of the Homestake Mine at Lead City, South Dakota around 1891. In 1876, Moses Manuel and his brother Fred struck what would turn out to be the richest single mine in the world. Three miles south of Deadwood they discovered a large vein of quartz (lead), which encased large amounts of gold. Using a water-powered *arrastra* that pulverized the quartz by dragging heavy boulders over it, the two mined over $5000 with their primitive operation, in a few months. After laying title to two other claims, they sold all three for $150,000 to investors Lloyd Tevis, James Haggin and George Hearst (father of the famous publisher). Hearst went on to spend $650 million mining the ore and with his aggressive business practices eventually took over $1 billion in gold from the Homestake Mines.

As in the other rushes, with the miners came many who had no intention of prospecting. They had their eyes set on making money without toiling in the mud and many made their fortune supplying the needs and wants of the prospectors, who had money to spend and weren't tight with it. In 1876, miners took $1,200,000 out of the Black Hills. At $20 an ounce, a fellow didn't mind paying $15 for a crate of eggs or $4 for the privilege of sleeping on a bed in a fancy hotel room. Businesses blossomed like spring flowers, giving quick and boisterous life to the towns that dotted the creeks of the Black Hills. While a miner could scratch an itch in any of a dozen or more of those towns, most of them eventually made their way to Deadwood, where they could do the scratching in a place that was becoming a legend.

Of all the places in the Black Hills brought to life by the gold rush, none reached the fame of Deadwood. Fixed at the junction of the Whitewood and Deadwood Creeks, the town came by its name honestly. When prospectors first stumbled upon the location, they could hardly make their way through the accumulation of fallen and rotting trees that strangled the site. Many have taken or been given credit for first discovering gold there; however, the man who most likely deserves the notoriety is Frank Bryant, who struck gold in August 1875. In the months that followed, more prospectors enjoyed similar success. During the first months of 1876, the most fortunate miners were washing upwards of $80,000 out of the standard 300-foot claim. If a fellow was lucky enough to hit a pocket of gold, $2000 per day wasn't out of the question. As often as not, however, he'd find that there were few such pockets on his claim.

The one thing that a gold strike does not abide is a secret. Through the fall and winter, newspapers were reporting that the Deadwood placer mines were the richest and most extensive ever discovered. Even cautious reports, such as those found in the respected *Harper's Weekly* suggesting that gold was only to be found in very low paying quantities, couldn't dampen enthusiasm. Given the amount of gold ultimately discovered, perhaps the best that can be said for the correspondent of *Harper's Weekly* is that he was probably somewhat misinformed. The fact was that reports of gold were often salted with plenty of exaggeration, much of it purposefully nurtured by the initial wave of prospectors who were anxious to sell some of the many claims they'd staked, at least those that weren't paying. *Scientific American*, for one, was onto their ploy, accurately informing its readers that Black Hills promoters were concocting "the most barefaced fabrications, got up by miners who wished to sell their claims." But, in such heady times, when optimism filled the air, folks were seduced by headlines and ignored the fine print. There were always willing buyers.

The tents that gave inadequate shelter to the stubborn who braved that first winter in Deadwood were rapidly replaced by wooden structures. It wasn't long before whipsawed pine buildings outnumbered roughly hewn log houses. On April 26, the newly elected municipal government laid out the town site and construction began in earnest. Despite plans there was little in the look of the town that had an organized appearance, which might be explained by the use of a compass and a lariat in place of proper surveying tools. The easy-going approach of the town founders was a good indication that they were less interested in civic-mindedness than they were in gaining federal recognition for the town. Accompanying such recognition was a federal grant of the

The town of Deadwood was about as bad as it got, a chaotic jumble of establishments thrown together within a single year. The haste in which this boomtown was constructed was matched only by the numbers of dissolute characters that flooded its streets. The large amounts of gold that were being found in the surrounding hills drew expert predators who smelled easy pickings. Rogues regularly held up stagecoaches, hoping to cash in on money being shipped in and out of town. Its saloons were frequently a veritable who's who of the most infamous gunslingers in the American West, with people such as Wyatt Earp, Wild Bill Hickok and Calamity Jane doing time in the nefarious burgh. There were almost 100 murders in Deadwood's first three years as a municipality, and the crime rate continued to rise as ever-increasing volumes of gold were being mined from the Black Hills. Things changed in the town when individual placer miners were replaced by large-scale operations. By late 1879, Deadwood's boom days were over.

land beneath the town site, title that increased real estate profits. Of course, the fly in the ointment was that the territory still belonged to the Sioux. Local officials were aware of that and took to calling their government "provisional." Federal officials looked the other way and, in so doing, justified the presence of General A.R.Z. Dawson, the local collector of federal revenues.

But the hodge-podge that was Deadwood was also the nature of a mining town in its infancy. It was a sight to behold, described by Frank Leslie's *Illustrated Magazine* as "one of the liveliest and queerest places west of the Mississippi." Travel down Main Street, which snaked along the canyon bottom, was as likely to be interrupted by a sluicing operation as it was a slow-moving wagon. Buildings and businesses mushroomed around the diggings so that tired miners would not have to go far to address their needs and wants. And in Deadwood, there were plenty anxious enough to take the edge off a hard day's work by using the amenities. Some estimates suggest that there were upwards of 25,000 residents in the summer of 1876, although a more realistic number was 10,000. Either number is staggering when it is remembered that it was a mess of brush half a year before.

Deadwood was rapidly transformed into a something more than a rough frontier town, though it certainly kept many of the edges associated with such a place. The lifeblood of the mining town was gold, but commerce was what kept it pumping. By fall there were over 170 businesses established, all reaching for the brass ring. There were hotels, drugstores, general stores, a theatre, and even a couple of lawyers threw up their shingles. Perhaps the best indication that there was money to burn was found in the fact that there were five peanut roasters in operation! Of course, prices for all goods were inflated and it wasn't just a result of the laws of supply and demand. Gold dust paid more than refined gold was worth at the time. Prices were accordingly marked up to cover the losses.

The town council was also active in rounding out services, most of which were supported by the revenue collected from licenses required by local businesses. The most significant institution to emerge was the local hospital. The council was active in making laws for the general betterment of the community in other ways. Citizens were apprised of the system for reporting contagious diseases, the proper disposal of refuse, appropriate sanitation practices, and the prohibition against the use of firearms. The rules and regulations were in place to ensure that Deadwood would become a model of respectability, and all that was required was for the locals to abide by them. It proved too much to ask. Truth be told, most officials weren't overly insistent on the matter anyway.

Side by side with the bank and the hospital were saloons and brothels, businesses that appealed to the physical appetites of the miners and made a good profit. Prospecting tended to attract men who weren't given to conventionality and moral stature. More often than not, they were running from the constraints of civilization as much as they were chasing gold. They were tough men, ready to swing both fist and pick. When the demanding physical business of work was done, they didn't give a second thought to unburdening the heavy pouches of their daily take. Men who live for the moment rarely do. A drink, a game of chance, a woman or all three was the only way many of them knew to relieve sore muscles and to recharge drained bodies.

More than one entrepreneur struck it rich at the bottom of a liquor barrel rather than at the bottom of a lode shaft. Saloon owners purchased a gallon of whisky for $1.65 in Cheyenne or Sioux City and sold it in Deadwood at $.50 for a one-ounce shot. A gallon brought in $64 and that was before the requisite watering down. Saloons enjoyed a booming business because, as miners were quick to say, theirs was a damn thirsty job. Ike Brown and Craven Lee were the first to turn that old adage into cash in Deadwood. Parched miners were so desperate to sooth their burning throats that they took to regularly frequenting the place, even though it was common enough knowledge that the swill Brown and Lee served up was more dangerous than the Sioux. Going down the gullet, it made veterans cross-eyed and it ensured that their stakes wouldn't be worked for the few days it took for its effects to wear off.

By the summer of 1877, there were 75 saloons in town. A miner could belly up to a plank nailed to a barrel at either end or, if his fortunes were better, he could go to a place with a carved bar, plate glass mirrors and music. As common as the shot glasses on the tables and just as popular, were the games of chance. The bigger saloons operated their own games; blackjack, faro and roulette quickly became favorite pastimes. Madame Moustache, who ran the faro bank at the Wide West Saloon was as well known as E.B. Farnum, the mayor, and perhaps even more popular. The fancier places attracted professional gamblers, out to lighten the pockets of those miners foolish enough to sit at a table with them. As most soon discovered, callused hands were no match for manicured fingers.

When a fellow sat down for a game of cards, he never knew who might be sitting across from him and, given the success of the card sharks who made good livings playing the game, most didn't seem to care either. Peering over his hand of cards, the miner might see a fashionable stranger puffing on an ivory cigarette holder or the cigar-chomping local madam, Poker Alice Tubbs, dressed in men's clothes and wearing an old army campaign hat. Occasionally he might even see a real celebrity, like James "Wild Bill" Hickock, Martha "Calamity Jane" Burke, or in later years, Wyatt Earp and Buffalo Bill Cody.

Wild Bill Hickock enjoyed a special connection with Deadwood. Throughout the Civil War, when he worked as a spy and scout for the Union forces, he had earned the reputation of having the fastest gun in the west. The standing served him well after the war when he made his living as a gambler. But on one August night in 1876, even he wasn't fast enough. Playing poker in Nutall and Mann's saloon, he was forced to take a seat with his back to the door, instead of to the corner (always the choice location of thoughtful gamblers). A fellow professional, Jack McCall, took the opportunity to fill his back with lead. As Hickock lay slumped over the table, his poker hand became visible for all to see. It was a pair of aces and a pair of eights, a pretty good hand, known ever since as the "Dead Man's Hand."

Gambling wasn't the only form of entertainment in Deadwood's saloons. There was music, some of it live. Helm's Union Park Brewery and Beer Gardens advertised a string and brass band and a spacious floor on which to kick up a heel. As often as not, the dancing was done by women only with cancans and the occasional striptease punctuating variety shows whose performers were best remembered for the vulgarity of their jokes. For women, the real money was to be made in offering services of a less public kind and they were none too shy about flaunting their wares. In a place where "irresponsible" women advertised in the town directory as well as on its street, there was obviously little to fear. The law generally looked the other way but if an "upstairs lady" from the Castle on Fink's Flat or the Gem was unfortunate enough to be hauled before the law, then she was likely to be released with the judge moaning about the

Martha Jane Burke (1852–1903), alias Calamity Jane, was a woman ahead of her time. She has been called frontier heroine, cowgirl and sharpshooter and claimed to be an army scout for Custer, wagon freighter and Indian fighter. However, it is possible that her most significant accomplishment was that she would not surrender to the establishment of the time, pursuing life on her own terms and being a role model for other free-spirited women. How she got her nickname remains a mystery. Some say it was because of her selfless caring for the sick during the smallpox epidemic, while others maintain that it came from the fact that women were often referred to as "Janes" and her penchant for attending shootings gained her the sobriquet, "calamity."

nefarious influence of the underworld. More likely to be punished were the men who used such services. If a fellow was brave enough to retire to a room with a Deadwood professional, then he had a good chance of leaving with only the clothes on his back, the rest being taken by the criminal element that associated with the ladies. If he was a respectable man (and many well-to-do gents frequented "questionable" saloons), he might wish he had been robbed rather than face the indignation of his wife. It was a challenge to keep such secrets in Deadwood and folks regularly chuckled at the sight of a fellow in a three-piece suit being chased down Main Street by his aggrieved better half.

By the early 1880s, when the heyday of the rush was over, efforts were underway to clean up Deadwood. Ordinances were passed to quash "tippling shops, billiard tables, ten-pin alleys, and ball alleys" as well as houses of prostitution. There was a crackdown on disorderliness, whether it was in the form of vagrants, beggars, gamblers or drunks. The new laws had all the impact of throwing a bucket of water on a forest fire. A testament to their ineffectiveness was the block of buildings on the west side of Main Street. In 1900, a fellow could climb to the second floor entrance of the structure on one end of the block, walk the distance to the other end, and witness a 20th-century manifestation of Sodom and Gomorrah with each step. As Buffalo Bill Cody put it around that time, "Deadwood was young so long that it will never quite forget its youth."

In the bonanza years of the Black Hills rush of '76, finding gold proved decidedly easier than getting it out of the mountains. Gold lured more than prospectors, businessmen, gamblers and dance hall girls. Where there was money, there were soon outlaws. In the early years of the rush, before the law had a chance to bring much order to the mining camps, trails and towns in the Black Hills, a man with a revolver and a willingness to use it pretty much had his way. He could strike gold as easily as a lucky prospector.

For many miners, violence was as much a part of life in the Black Hills as was prospecting. Initially, the threats came from the Sioux or the Cheyenne who were intent on "prosecuting" trespassers, and doing it in a way that would send a clear message to others that they weren't welcome in the Black Hills. Throughout 1875 and 1876, the trails leading into gold country were bloodied with the mutilated bodies of men who had dreamed of gold but met with the nightmare of Native anger. Residents of Deadwood developed their own strategy for dealing with that. Officials voted a reward of $25 for each Native brought in, dead or alive. It eventually jumped to $250 and all a fellow needed to collect his due was a head. The local board of health was even known to make payouts by rationalizing that killing Natives was good for the health of the community! Officials might have better served their townsfolk by looking to the criminal element among the white population. Even before the Sioux had relinquished the Black Hills, white outlaws were in on the action. They teamed up with the Natives, stoked their rage, divvied up the loot and left their "allies" to shoulder the blame.

The most infamous example of such an alliance occurred in the spring of 1876, when the Metz party was wiped out. Metz was a baker out of Custer. As he watched the stream of prospectors headed to the Black Hills, he found that he, too, was seduced by temptation. Selling his business, he threw all his money into his prospecting operation and joined the current flowing north. By the time he left Cheyenne there were seven in the party, including his wife and a maid. The party became separated in the Red Canyon, in the southern part of the mountains. Natives attacked the three at the rear. "California Bill" Felton received five wounds, but still was able to load his two injured friends on horses and lead them to a nearby stage station. Likely assuming the three were dead, the Natives continued up the valley where they found the remainder of the party. They killed Metz, his wife, the maid and a companion and left their bodies on the trail as a grisly statement to other would-be prospectors. Felton was the only member of the party to survive. When the carnage was cleared, tracks suggested that white outlaws had a hand in the attack. Rumor had it that "Persimmions Bill," a bandit well known in the Red Canyon area, made off with a hefty share of the spoils.

Outlawry's glory days paralleled the growth of Deadwood. By the summer of 1876, when the town had emerged as dominant in the Black Hills, stagecoach routes snaked out in all directions as surrounding cities sought to cash in on the bonanza. They wanted prospectors to filter through their streets, spend money in their hotels and bars, and buy supplies from their shopkeepers. For that to happen, they needed a good link to the Black Hills. Soon there were direct routes to Cheyenne and Bismarck and connecting routes to Sidney and Fort Pierre. Freighters who charged upwards of $5 for 100 pounds made good money transporting goods. So did the bandits who preyed on travelers.

For the most part, outlaws ignored the freighters and with good reason. A haul of cargo was pretty much useless until it was turned into cash. That meant stealing the cargo and then the added burdens of transporting the goods and selling them. The market for most cargo was limited in the

Black Hills and only the most naive wouldn't know that a fellow met on the trail peddling goods had stolen them somewhere along the way. No, the real money was to be made robbing stagecoaches heading out of the Black Hills because they carried the gold.

In the early days of the rush, it was common for people to carry their valuables with them on passenger coaches. Usually they were safe enough around Deadwood, but when they traveled towards the empty mountain fringes, the journey got dicey. Passenger coach holdups became so common that most folks came to look on it as part of doing business in the Black Hills. Some took to fighting the bandits and, while they were occasionally successful, more often than not they'd still lose their gold and occasionally their lives. As outlaws like "Persimmons Bill" and Sam Bass got richer, folks learned that passenger coaches were not the best way to transport their riches. Coach lines took to demanding that customers turn their valuables into non-cashable bonds before they'd let them ride. Others developed cunning strategies designed to fool their assailants.

Sometimes this meant finding a hideaway inside the coach, but a practiced bandit could strip down a vehicle in less time than it took to hold it up. More interesting were the efforts of passengers to conceal goods on their persons. As a result of the passengers' industry, there wasn't a self-respecting bandit in the Black Hills who didn't have a doctor's intimate knowledge of the shape of a human body. Clothes were carefully rifled and uncharacteristic bumps were carefully investigated. Even then, there were occasions when would-be victims would use the skills at their disposal to prevent a theft. One of Deadwood's "upstairs ladies" wound her money into her hair in a tight bun. When she willingly allowed the bandits to proceed with their search, they welcomed their task with such enthusiasm that they didn't even think of searching

The Black Hills

The stagecoach was the most common mode of public transportation before the coming of the railway for residents of the West and for miners leaving the Black Hills with their gold. The name is derived from the fixed points along the route where the coaches stopped to change horses, pick up mail, provide relief for weary passengers and deliver freight. These coaches were made of wood and steel, typically weighed in at around 3000 pounds and were drawn by either four or six horses matched in teams of two and controlled by the driver with one set of reins. The interior of a coach often had three bench seats and carried up to nine passengers, each of whom could take only 25 pounds of luggage. The saying "riding shotgun" may have its origins in this mode of transportation since it was necessary for the driver to be partnered with an armed guard who could protect the coach, its valuable cargo and its passengers from bandits.

above her shoulders. Another passenger was greeted with the outlaw's command of "hands up." When he objected to that "high-handed piece of business," the thieves thought his comeback so amusing that they let him keep his watch!

Most in the Black Hills eventually came to seek safer ways to get their earnings out of mountains. Coach owners soon realized that there was a buck to be made in facilitating safer travel, so they brought in stagecoaches, which didn't usually carry passengers, to run the routes. It took some time for them to develop a system that could foil the efforts of determined outlaws. From the beginning, stagecoaches were an easy target. Lonely drivers traveled long distances over isolated trails. And when, as driver Tom Cooper complained, a rig is regularly overtaken by bull teams, mud turtles and men in wheelchairs, it's clear enough that outlaws had an easy enough job bringing the stagecoaches under control. Getting the gold was no more difficult. Initially, it was placed in steel strong boxes so it was easy enough to ride off with the treasure. And, while they were always secured with an imposing lock, the gold inside was only a bullet away—the device that could withstand a gunshot was rare indeed. Soon outlaws discovered that operators had taken to anchoring the strong boxes to the coach. It was no deterrent because it only meant that they had to shoot the lock off in the presence of the driver. It wasn't until the coaching outfits added guards and armored plates that bandits faced a challenge. And a challenge it was! If a bandit got too close to a coach, he was likely to be brought down by the man riding shotgun. If he preferred to try his luck by firing at the stagecoach from the shelter of the trees, he discovered that his bullets hardly dented the one-inch thick steel plates. Despite these precautions, neither the guard nor the steel put much of a dent in the outlaws' activities.

It's difficult to know exactly how much gold was netted in holdups. As folks began to abandon passenger coaches as a way to get their gold out of the Black Hills, the prizes taken in those operations increasingly became little more than a few dollars. The stagecoaches were a different matter because they carried riches, or at least were rumored to, which made it worth putting one's life on the line. The fact was that stagecoach operators didn't want the amount of gold they carried to become common knowledge for fear of drawing an even bigger bull's eye on the side of their four-wheelers. Nevertheless, hints as to the amount of gold carried by the stagecoaches did occasionally surface. Sometimes thefts were so large that owners felt it necessary to post a reward for its return. One such case involved a so called "treasure coach" out of the Homestake. That company's coaches had long drawn greedy glances from bandits, but most eventually turned away because they were simply too well guarded. Charles Carey and his four-man gang decided to use a little creativity in their approach. They figured the risk was worth it because rumor had it that the coach carried some $250,000 worth of gold.

Instead of attacking the coach on the trail, they descended on the coach station at Canyon Springs. Poor old "Stuttering Dick" Wright, the stock tender, was tied up before he knew what was going on. When the coach arrived, the usual cup of hot tea that awaited them was replaced by a helping of hot lead. In the first volley, one of the guards was killed and the other two were wounded. One of the injured, Scott Davis, had the strength to continue to return fire, but when Carey grabbed the driver and used him as a shield to advance on Davis, the guard fled for help. With the guards out of the picture, the Carey gang turned to the chest. Forcing it open, they discovered $27,000 worth of gold, a disappointment, but still a respectable take. They packed up the gold and wasted no more time before hightailing it. The body of one of their companions, Big Nose George, was left where he fell. Soon a posse was on their trail. While they were camped near Rapid City, it nearly caught them. Surprised, the gang hurriedly left its diggings, lighter by an $11,000 gold bar. In the months that followed, most of the gold was recovered and two of the gang members were arrested and thrown in jail.

The outlaws who found themselves in the hoosegow for their ways were often the lucky ones. Cornelius "Lame Johnny" Donahue roamed along the eastern foothills, preying on those coaches that took the Sidney route. He was eventually captured and, while being transported to Deadwood for trial, met *his* end at the end of a rope provided by philanthropic vigilantes. His friends later discovered his grave and erected a proper memorial.

> *Pilgrim, pause.*
> *You're standing on the mouldering bones of Limping John.*
> *Tread Lightly, stranger, upon this sod,*
> *For if you make a move, by God, you're robbed!*

Deadwood, too, enjoyed its share of robbery and murder. Cheyenne newspapers suggested that the town might want to change its name to Deadman, with good cause since there were nearly 100 killings in the vicinity from 1876 to 1879. The number shouldn't be surprising. Shocked visitors to the town described every man as carrying 14 pounds of firearms and possessing a willingness to use them before words got in the way. Seth Bullock, a pioneer in the Black Hills and one of its first sheriffs, did as good a job as he could of rounding up criminals. Those who slipped through his fingers likely regretted it because they weren't as likely to slip through the neck-tie of hastily formed lynching parties. Even honest folks had to be concerned about the enthusiasm of vigilantes since it wasn't unheard of for them to hang the wrong fellow. The errant lawmen took such unfortunate accidents in stride. They were only enforcing the will of the community so even the death of an innocent man was believed to give would-be criminals pause.

Lesser crimes were so widespread in Deadwood that the papers soon stopped reporting them. There was more theft than there was gold dust. If it wasn't nailed down, it had a good chance of disappearing and folks kept Bullock busy with sad stories of missing chickens, cordwood and food supplies. Tools regularly disappeared, and if a miner wrapped up his day's work before emptying his sluice boxes of amalgam, he had a good chance of finding that an unknown friend had finished the job for him by moonlight. Desperate thieves soon learned miners had no second thoughts about using their rifles to protect even their picks. In a place where crime was as common as dreams of gold, it should have come as no surprise.

ഇ

By 1879 the Black Hills rush was in full swing and gold was being produced at a rate of $300,000 month, somewhat more than was mined in the first three years of the rush, which estimates place at anywhere between $6 to $10 million. But the nature of the game had changed. No longer were significant amounts panned from placers. It had been a few years since a fellow could even stake out one of those operations. More and more, prospectors turned to lode claims and hard-rock mining. Sinking shafts, using hydraulic equipment and building specialized mills to extract the gold from the ore was an expensive undertaking and a sure sign that the rush was maturing. Father De Smet sold his mine for $400,000 and others collected similar amounts, clear indications that investment capital was required and that there was still good money to be made. Men whose interest was a daily wage rather than the excitement of prospecting worked hard-rock mines. Only three years into the rush, the day when a man could buy his kit and head off for riches was already a thing of the past.

The population around the Black Hills dropped off through 1879 (it probably wasn't more than 20,000) though it was revitalized with the news of hard-rock mining. A few new towns even sprang to life. However, the wild-eyed enthusiasm of the early days was gone, replaced by the resignation that accompanied a working life below ground. But as much as anything else, it was an accident that dropped the curtain on the Black Hills bonanza. On the evening of September 25, 1879, a fire broke out in Mrs. Ellsner's Star Bakery. Whipped by the winds of the Whitewood valley, the flames quickly leaped to Jensen and Bliss's hardware store. By then, residents were cursing the shortsightedness that had led them to dismiss calls for fire fighting equipment. When the fire exploded eight kegs of gunpowder launching hot embers over the town, residents accepted the fact that there was little to do but watch the brilliant blaze dance against the night sky. When the flames died out, 300 buildings were charred rubble and 40 percent of the population was homeless. Estimates of loss reached $3,000,000.

The town rebuilt, but it was merchants and not prospectors who were behind the effort. The rush was over. An occasional visitor could still be found hiking through the mountains. He'd be pretty safe because the Sioux were on their reservations and the outlaws were working more well-known trails. And he'd be driven by the hope that he'd have the luck of William Franklin, who discovered the Holy Terror mine while digging a posthole. Named for his wife, it produced $100,000 in a few weeks. On such stories gold rushes were built.

The Klondike
(1897)

"Skagus," thought William Moore. "Skagus...the home of the North Wind...." He turned up the collar of his black woolen coat for some protection against the wind and gave his arms a good, brisk rub. "Those red men might just be right at that," he concluded. Moore stood at the end of a wharf that stretched out a mile over the tidal flats of Skagway Bay. Just beyond was Lynn Canal that opened into the northern reaches of the puzzle of islands that formed a good chunk of the Alaska panhandle. Behind him, the winds continued to tumble along until the Coast Mountains eventually brought them to a sudden halt. There was one spot, however, where the winds could continue unabated. It was a narrow channel located at the southern end of Summit Lake where the Skagway River had carved its route to the bay. When Moore first saw that pass, his heart fluttered. When he traveled through it and learned that it was deep enough for the easy transit of both man and beast, his heart raced. Moore sensed that the place was destined for great things.

The wind was watering up Moore's eyes and he blinked to clear them. He looked out across the bay and chewed on the well-pitted stem of a pipe. Neighbors might have thought him crazy had there been any. Moore lived with his wife in a cabin near the lowland where the river emptied into the harbor, and their closest acquaintance was two bays away over at Dyea. He knew Alaska well, but he hadn't arrived in Skagway until 1887 soon after his son had sent him word about the pass. From the first, he was excited about the news and he wasted no time in sailing north to give the place a once over. It didn't take much more than a single glance for Moore to stake his claim, but he didn't do it with the intention of mining the Skagway. Moore was convinced that significant gold deposits were hidden in the northern interior. He saw his fortune in selling the land in the cove and around the river to businessmen anxious to provision those who would be swept along in the inevitable rush.

There was some reason for Moore's optimism. There had been a few gold strikes in the interior by the 1890s, but it had been nothing substantial. What there was plenty of, however, was Moore's determination. He often told his sons that he envisioned heavily laden steamers docking at his wharf, with passengers and supplies filling the city that was sure to take shape. "Skagway is a place with a future," he'd confidently bark, "and I'm going to have an important place in it." Moore wasn't about to just sit idly by and wait for it to happen. A few years back, he had surveyed the pass with the help of Skookum Jim Mason who was to enjoy Klondike fame in his own right. He brought back the evidence that was to put Skagway on the map. At 33 miles, the White Pass (Moore had named it after Sir Thomas White, the Canadian Minister of the Interior) was some five miles longer than the nearby Chilkoot Trail from Dyea, but it was

1000 feet lower and much easier to travel. It was good news, so on the basis of it, he spent countless hours surveying the town site and dividing it into lots.

For the better part of a decade, he hadn't wavered in his faith. Finally there was news of a big strike on the Klondike River. As he stood on the wharf and yet again imagined his city, Moore felt the hairs on his neck prickle. There seemed to be something in the air. It was the spring of 1897 and he sensed that his dreams were about to be realized. But it was more than intuition.

Moore knew something of gold and the way it made men's blood run hot. He had seen enough of gold rushes to know that there was usually good reason to act on a hunch. He was a salty dog in his 70s and had spent most of the previous half century following rush after rush—California, Peru, the Fraser River, the Cariboo. He had seen fortunes made and lost, one of which was his own earned in the steamship business. The one thing he'd learned from his journeys was that mining was an uncertain business. Striking it rich in the gold fields had as much to do with being born under the right star as it did with hard work. But there *was* easy money to be made, money that was never in doubt. If a fellow could sell what the miners wanted, he had a license to print money. In Skagway, Moore had a monopoly on that which was most valuable: land. Yet, for all his experience, it turned out that Moore was totally unprepared for what was to come.

Moore was right about the White Pass, too right as it turned out. It did attract miners seeking the easiest and quickest route to the gold fields. The *Queen*, the first of the Klondike stampeders' steamers, slipped into Skagway Bay on July 26, 1897. With it Moore's dreams evaporated. Tranquillity was immediately transformed into chaos. Men, unceremoniously dumped on the beach, rushed up the

Moore's Wharf at Skagway, 1904. Captain William Moore and his son, J. Bernard, were the first white settlers to make their home in Skagway. When Moore became the first European to cross what he named the White Pass, he discovered a natural route from tidewater to the interior lakes and laid claim to 160 acres in the Skagway River Valley in 1887. He predicted accurately that when gold was discovered along the Klondike River, the White Pass would be one of the few routes that could provide relatively reasonable access to the Yukon gold fields. He sought profit from the prospectors' use of his land, but was thwarted in his attempts. He had better luck charging fees for the use of his wharf, but was forced to share profits with outside investors because he lacked the necessary funds to expand and meet the needs of the hordes of prospectors. By 1898, three other companies had built wharves. Moore came out on top, however, when the railroad finally made its way to Skagway, because his wharf was closest to the tracks.

Skagway sits atop the Inside Passage, nestled between Taiya Inlet and the Coast Mountains. Its name descends from the Tlingit word *skagus*, which means "home of the north wind." First settled by William Moore, Skagway profited greatly when gold was discovered in the Klondike region in July 1897; thousands flocked there because of its proximity to the White Pass. The path proved treacherous; in the winter of 1897–98, owners drove 3000 pack animals to their deaths, and the White Pass became more colorfully known as Dead Horse Trail. While the Chilkoot Trail from Dyea became the route of choice, Skagway's booming population of 20,000 made it the largest town in Alaska for its time. Many gold rush towns were lawless, but Skagway had the dubious honor of being "the roughest place in the world," at least in the opinion of Superintendent Sam Steele of the NWMP. Home to 80 saloons, outlaw gangs, and a steady stream of painted women, many were drawn to Skagway not by gold, but by the entrepreneurial spirit that was a fitting tribute to the town's founder, William Moore.

shore and shoved by Moore, ignoring his claims that the land was his. They weren't for taking directions and they were certainly not interested in keeping things orderly, taking the spots they wanted and hacking away at the trees to build makeshift shelters. Those too impatient took to burning them down. By August, the numbers of newcomers had grown enough to allow for a local town government. One of its first acts was to survey the town. Moore pointed out that he had already done the legwork and he pushed for developments along the lines of his town site plans. Baffled newcomers merely looked at him as if he were some strange old codger, too long living in isolation. Lot registration fees were collected and city coffers swelled with the money that Moore felt was properly destined for his pockets.

The final humiliation came with the report that one of the newly surveyed roads took a course right through Moore's house. He was being evicted! Moore met the order to move armed with a crowbar. The evictors arrived with handspikes. The confrontation only proved that Moore was more determined to use his weapon than were the city officials, so the crowd dispersed. Moore could see what was in the cards so he soon packed up and moved to another part of town. As much of a nightmare as his dream had become, he wasn't going to be kicked out of town. For one thing, he was making a fortune from rental fees for the use of his wharf. And he hadn't given up on what he saw as his land rights. A tenacious Moore took the matter to court. After four years, a judgement was made in his favor. He was to receive 25% of the assessed value of all the lots that lay within his original town site. Though he witnessed the utter destruction of his magnificent retreat, he died a rich and wiser man. No one can control a gold rush.

History is two-faced in the northern interior that was to boom as the Klondike after the discovery of gold in 1896. One set of eyes peers into the misty beginnings of a human presence in North America. Sharp enough sight reveals a copper-toned people. Science suggests that they were travelers from Asia, a nomadic people following the animal herds across Beringia, the land mass that temporarily linked the two continents. However, the people themselves—the Tlingit, Tutchone, Hän and others—gather their children around twilight fires and tell the legends of their beginnings, legends that fix their origin on Turtle Island. They did not come to North America, it is explained. They were created here.

The second pair of eyes do not need to strain in their search for the origin of a white presence. It could be found in the same century that the gold rush occurred. The white men who opened what were to become Alaska and the Yukon came primarily by sea. The Russians were the first formidable presence there. In 1741 the great explorer Vitus Bering discovered Alaska. The name is derived from the Innu word *Alakshah*, meaning "mainland." Bering's ship was wrecked on the voyage and he didn't return to the motherland. However, a ship full of sea otter pelts, likely obtained in trade from the coastal Tlingit, did. While it is possible that Russian *promyshlenniki* (fur traders) were active in the region prior to this date, it was Bering's discovery that caught Russian imaginations, if not their ledger books. For 60 years the *promyshlenniki* plied their trade as independent operators.

At the beginning of the 19th century, Russia decided to formalize its trading efforts in the region and so created the Russian American Company. It was as much a political move as an economic one, designed to ward off the northern expansion of the Spanish and the British, the latter of whom showed particular interest in the territories of the Northwest Pacific. James Cook had sailed around Alaska in the late 1780s and George Vancouver had explored what would become known as the Alaskan panhandle in 1792. The Russian American Company was not long in operation when gold first made its appearance. Around the turn of the century at Sitka, the seat of the Russian governor, Alexander Baranov, a trader showed up with 50 gold nuggets allegedly taken from an inland river. Baranov did not want news of the discovery to spread; he enjoyed too much the fine trappings of his colonial court to have it disrupted by a horde of gold seekers. Legend enriches the tale by suggesting that Baranov shot the messenger to ensure that tales of gold would not spread. Whatever he did, the find was effectively kept secret.

Russia's hold on Alaska was not to last. Through the first half of the 19th century, the fortunes of both the colony and the motherland suffered. Alaska faced increasing competition from other nations active in the northern Pacific, including the Americans. Russia tried to protect her interests by establishing missions and isolating the colony. It forbade trade with foreign nations and it signed a treaty with the British abandoning all claims south of 54° 40'. The same treaty set the inland boundary of Russian sovereignty at the 141st meridian (though the coastal boundary remained indistinct, described as being at the first range of mountains). The decrees against foreign trade were disastrous because the colony depended on such business for its survival. More important to Alaska's future, however, were Russia's own fortunes. The Crimean War began in 1853, and within four years the Russian ambassador in Washington was whispering to American officials of the cash-strapped Czar's desire to sell Alaska. In 1859 negotiations began. They were completed in 1867. The delay was a result of the American Civil War and not indicative of American indifference. The purchase price was $7.2 million.

The exploration of the interior waited until the latter part of the 18th century and even then it was slow going. The British were the first to enter the region as fur traders associated with the North West Company. They were seeking out new trading partners and a functional route to the Pacific that would allow them to avoid the great expenses of shipping furs back to ports in eastern British North America. Alexander Mackenzie set out from Great Slave Lake in 1789 and followed the river, which today bears his name, to the Arctic. It's probable that he was the first white man that the Athapaskans of the western subarctic had ever seen. Throughout the middle decades of the 19th century several trading posts opened in the western reaches of the North-West Territories, but it was the mid-1870s before independent operators built the first such establishment, Fort Reliance, near what would be gold rush country.

The last quarter of the 19th century finally saw men arrive in search of gold. In 1873 a gold strike in the Cassiar Mountains, which straddled British Columbia and what would soon be known as the Yukon, enjoyed brief notoriety. When it petered out, the dispersing miners mostly made for the coast, where some of them continued prospecting. In 1880 Joe Juneau stumbled upon rich placer deposits in the Silver Bow Basin. Within a year, the boom town of Juneau had sprung to life but it only produced about half a million dollars in gold. French Pete Erussard also discovered a rich vein nearby. None of these strikes were big when compared to the discoveries that had set off the crazed rushes in the south, but they were sufficient to allow the region to gain a reputation as gold country.

It's in the nature of prospectors to be drawn to places where such talk is heard, especially when there seemed to be something to it. For those who thought about it, gold in the upper reaches of North America made good sense. If the yellow ore was found in California and British Columbia, then it was logical to assume that it would be found along that same line farther to the north. Such reasoning brought Arthur Harper into the region in 1873. Harper had ridden the rushes along the Fraser River and in the Cariboo, so he was no greenhorn. He followed a route similar to Mackenzie's from the southern interior along the great river and finally the Porcupine (an old fur trading trail). For 25 years he explored the Mackenzie River tributaries and the streams that slipped through the Mackenzie Mountains. He went west to the Yukon River. He even panned on the Klondike, but found no gold. He died in 1897, blessed with the knowledge of the discovery of gold but cursed with a sick body too weak to join in the rush.

George Holt picked his way through the Coast Mountains in 1878; he was probably the first white man to do so. It was no mean feat. His journey took him through the Chilkoot Pass, one of the few flaws in an otherwise secure mountain fortress. The challenge of climbing through the pass was substantial enough, but success in that endeavor only meant that Holt had to face a more formidable enemy, the Chilkoot, who guarded the pass. They knew that the pass gave access to the Yukon Valley and inland trade and they were determined to keep intruders out. The stars were aligned for Holt because he was able to scamper into the interior and return to the coast with a few gold nuggets. When folks in Sitka saw the gleaming evidence, they were determined to follow Holt's footsteps. The Chilkoot were even more vigilant on this occasion, but the prospectors arrived under the protection of the United States gunboat S.S. *Jamestown*. The Natives proved receptive after an appropriate number of rounds fired from a Gatling gun. But all was not lost for the Chilkoot because they were smart traders and began charging the newcomers to carry their supplies. When the gold rush broke, they were making a very respectable $1 a pound. They weren't carrying Holt's supplies though. He died at Native hands farther inland without ever having struck it rich.

Other prospectors followed. In the early 1880s Ed Schieffelin, who had set the Tombstone silver rush in motion, stole into the Yukon River by way of the Bering Sea. He found some gold, but by the time he did, it was late in the season and he had to retreat without extensive mining. Jack McQuesten and Al Mayo joined with Harper and established Fort Reliance and, later, a string of smaller posts along the Yukon. They did some prospecting, but their real contribution to the rush was in providing supplies that sustained the slow trickle of miners into the region, of whom there were some 200 by the mid-1880s. The small rush town of Fortymile emerged just east of the Alaska/Yukon boundary on the Yukon River. A half dozen years later, there was a strike at Sixtymile (in the other direction along the Yukon River). Another strike quickly followed in what would soon be called Circle City, some 240 miles down the Yukon River in Alaska. It wasn't until 1896 that prospectors hit upon the discovery that opened the flood gates to the rush of '97.

Robert Henderson was one of those pulled north by the news of a string of small finds. Born in Nova Scotia, he had left home as a young man to live his dreams as a prospector. He had been in California, Australia and New Zealand, always driven by an unwavering goal—he wanted to hit the big strike. In 1896, he was in Ogilvie on the Fortymile

For years, the Chilkoot people had used this 32-mile trail to cross from the Yukon River valley to the Pacific coast. Along with the White Pass, the Chilkoot Pass was one of the only land routes through the Coast Mountains. When gold was discovered in the Klondike in 1897, the Chilkoot became a crucial passage to the gold fields. The pass leads from Dyea to the shores of Lake Bennett, where prospectors would then travel by water to Dawson City. While most of the trail was suitable for walking, the pass itself was a cruel 3700 feet of steep mountainside. It was easiest to traverse in the winter when snow and ice were fashioned into the "Golden Stairs." The Canadian government didn't make things easier. Once prospectors reached the summit, the NWMP would only let them enter Canada if they carried enough food supplies for a year. This meant that miners climbed the Chilkoot again and again for three months, hauling up endless loads from their 1000- pound cache of food.

River. Up to that point, he had found precious little in the way of gold and he was running low on supplies. There was some luck in all the major strikes, and on this occasion it came in the form of Joseph Ladue, a local trader. He offered to grubstake Henderson, which meant he'd give the prospector supplies in return for a cut of what he found. Ladue saw it as a way to drum up business. If Henderson did discover gold, Ladue would have his own gold mine in the form of his trading post. Henderson eventually ended up on a tributary of the Klondike (a white man's corruption of the Native name for the river *Tron-Dec*), where his panning paid $.08 a go. It was a little less than what was considered to be the norm of $.10 a pan. Hopeful nevertheless, he named the stream Gold Bottom.

Henderson returned to Ladue's to stock up when he encountered George Carmack. The son of a forty-niner, Carmack had been in the region for about 10 years. He wasn't so much interested in gold. He was trying to escape the trappings of an increasingly commercial and technological American society by searching for a way of life more in tune with nature. To that end he had married a Tagish woman and lived with the Natives. Henderson wasn't secretive about his find and Carmack was one of the folks with whom he shared it. Most were uninterested in what was at best an average strike, but Carmack had time on his hands and decided to investigate. Skookum Jim and Tagish Charley, relatives of his wife, accompanied him. After some panning, they decided that Gold Bottom was about as misleading a name as ever given to a stream. Carmack was ready to head back to his wife when Tagish Charley suggested they give nearby Rabbit Creek a try. That suggestion shouldered the Klondike Gold Rush.

Who found the gold is a mystery. Carmack claimed he did, but Skookum Jim and Tagish Charley asserted that

George Carmack (seated at left) and his Native brothers-in-law, Skookum Jim (second from the right) and Tagish Charlie (not pictured), were credited with finding gold in 1896 at Rabbit Creek, which was quickly renamed Bonanza Creek, setting off the Klondike Gold Rush. The great irony is that Carmack wasn't even interested in gold. Being the son of a forty-niner, he knew of the trials and tribulations endured by most prospectors. As fate would have it, he made the first claim on Bonanza Creek. Pictured here with the two prospectors are Skookum Jim's wife and daughter (center) and two unidentified men. The village of Carmacks in the Yukon Territory at the mouth of the Nordenskiold River is also named for George Carmack because he built a cabin at that location in 1893. The town became a supply depot for miners and riverboats in the era of the gold rush.

Carmack was asleep when, on August 17, 1896, Skookum Jim pulled two golden nuggets from the stream. If he was in slumber, the excitement surely woke him up. They took to their pans and were soon pulling out $4 of gold with every scoop of gravel. They were so excited that they smoked all their cigarettes in celebration! The discovery took them totally unawares. They had nothing to carry the gold dust in. A desperate Carmack rifled through his possessions until his eyes fell on a round of ammunition. He fired it off and filled the shell with gold. Before they left, the trio staked their claims. Carmack took two claims of 500 feet each and, since only the man who discovered gold was permitted two claims, his 1000 feet might solve the mystery of who actually found the nuggets. Skookum Jim's and Tagish Charley's claims were identified by their position relative to Carmack's original claim, One Above and Two Below, respectively.

When the trio returned to Fortymile to register their claims, they had trouble finding anyone who would believe them. The gold displayed as evidence was considered to be of a texture and color alien to the north, so some claimed that a con was on. Others weren't willing to take such a chance. Inspector Constantine, commanding officer of the Yukon District of the North-West Mounted Police, summed up the subsequent rush in a report to his superiors, "Men left their old claims and with a blanket, axe and a few hardtack prospected on the new creek [soon dubbed Bonanza], staked, and registered their claims which in all cases gave better prospects than any other heretofore." Within two weeks there were some 200 claims. Too many for the Bonanza Creek, the miners spilled onto the Eldorado, an even richer tributary or "pup." Ladue smelled something big and staked out a town site at the junction of the Klondike and Yukon rivers. He called it Dawson, after the Canadian geologist George Dawson. He then built a warehouse and a sawmill for the construction boom he sensed was inevitable. And they did come. It was slow going at first as the news of $200 pans ambled to Circle City. By the summer of 1897, this first wave of the rush had crested at about 4000. When news climbed the Coast Mountains and eventually reached the world at large, there would be a wave of tidal proportions previously not witnessed in North America.

As for the five men involved in the discovery, they mostly did fairly well. Skookum Jim continued to prospect and kept a share in One Above, a share that netted him $90,000 a year until he died in 1916. Tagish Charley sold out in 1901 and used his profits to open a restaurant in Carcross. But the bottle proved too magnetic to him and one night he drowned while drunk. Money changed Carmack's view on the romantic attractions of nature. He, too, sold his share and left the Klondike and his wife for California, where he opened an apartment and hotel. The Canadian government eventually recognized Henderson as the legitimate discoverer of Klondike gold and he was granted a pension of $200 a month, the only money he ever saw as a result of the discovery. The fact that he was a Canadian undoubtedly influenced the government's decision. He died in 1933, still searching for a strike he could mine. And Joseph Ladue made a fortune—he was worth $5 million when he died—which is hardly surprising. He was, after all, the shopkeeper.

༄༅

Throughout late 1896 and 1897, reports of the Klondike strike had trickled down the coast, but the news had lost most of its steam by the time it reached the big cities. At best, newspapers devoted a few speculative lines to the northern digs. It wouldn't be long before they were writing about little else.

On July 14, a small, rusty steamship named the *Excelsior* chugged into San Francisco Bay. Her arrival would surely have gone without comment or notice save for the fact she had come down from the north and was rumored to have Klondike miners aboard. A handful of curious folks, those few who had followed the sporadic reports about gold in the north, were waiting on the dock. As they waited, most were loudly debating the merits of the reports. The issue would soon enough be resolved. The onlookers fell silent as they caught sight of the vessel's unkempt passengers—bushy beards, dark faces, ragged and patched clothing, slouch hats that had lost what form they once had. There was no doubt that they were miners. But had they struck it rich? As the eyes of the onlookers fell upon suitcases, canvas sacks and buckskin bags so heavy that many of them required two men to lift them, their question was answered.

The miners hailed a four-horse truck and ordered it to head to the United States Mint. It was closed that day, so they went instead to Selby Smelting Works. As the truck wound its way through the streets of San Francisco, the crowd behind it multiplied and even a few grizzled forty-niners were seen hobbling along at the rear. Soon they were at Selby's and astonished bystanders watched as the sacks were emptied. One miner had $35,000 worth of gold, another $50,000. A less fortunate soul could only produce $15,000, but with whisky at $.40 a quart and a fancy apartment at $1.25 a week, even *that* could be stretched pretty far. One punchy fellow ordered a $20-a-day horse cab for a fortnight exclaiming, "By God, I pulled a hand sled 1400 miles and now I intend on riding in luxury!"

The next day the scene was repeated in Seattle with the arrival of the *Portland*. The only difference was that the

Excelsior's cargo had heightened interest in the arrival of the *Portland*, that and the fact that the *Portland* reportedly carried enough gold to put to shame what was on board the *Excelsior*. Reporters were ferried out to the ship before it steamed into the sound and were shown such sights that most had only imagined. Blinded by the glare of the gold, they stumbled back into their tugs and were onshore writing the stories within minutes. By the time the *Portland* docked, newspapers were being flogged. Headlines boomed "Gold!" and "Stacks of Yellow Metal!," but it was the story in the *Post-Intelligencer* that captured the imagination. The *Portland* carried "more than a ton of gold." In fact, it carried about 2 tons, and 5000 residents of Seattle watched the miners unload it in the pre-dawn hours. The news from Seattle and San Francisco, reporting a total of $1.5 million in gold, was quickly reprinted in newspapers across the country and then the world. The Klondike stampede was on.

The 4000 who called the Klondike home in the summer of 1897 increased ten-fold over the next two years. The first wave following the news of the *Excelsior* and the *Portland* was relatively small. With an early winter closing in, most believed that the window of opportunity was but a few months wide, so only the most experienced and determined of the 10,000 who set out were settled in before the snow fell. The final and largest wave arrived throughout 1898, with some spilling into 1899. Many of the 40,000 ended up in Dawson City, sending it into a crazed boom the likes of which even gold rush towns like Deadwood and Barkerville had not witnessed. The singular enthusiasm that characterized the Klondike rush was the stagnant economy of the 1890s. Wall Street had crashed in 1893, and that had been followed by a depression, which was soon an international affair. Klondike gold stimulated trade, whetted the demand for services and sent economies racing. Of course, the vagaries of international economics were the farthest thing from the minds of those who made tracks north. With the cry, "Hurrah for the Klondike!" some hundred thousand would-be miners around the world said their goodbyes to families. That fewer than half of them actually arrived had something to do with the challenges of getting there and a general ignorance of those challenges.

It's an understatement to suggest that those bound for the Klondike knew little of the place. At most, they could say it was to the north, the same as California was to the west. But it wasn't as easy to get to as was California. No trains chugged north as some counted on, and those who purchased bicycles for the trip were forced to return them or abandon them along the way. The obstacles involved in getting to the place undoubtedly deterred a great number of the million or so folks who investigated traveling north. The diligent discovered that Dawson was accessible by some eight routes, though accessible was apparently a flexible term. Would-be Kondikers were at the mercy of those who profited by transporting people into the region. Occasionally colorful pamphlets obscured the motives of unscrupulous promoters by concealing the somber reality of the advertised route, as is evident in a series of reports printed in *The Klondike News* in 1898. "We warn our readers against any attempt to reach the Klondike country by way of Copper River. No living man ever made the trip, and the bones of many a prospector whiten the way." As for the Takau Route, it "is another back-breaking, soul-destroying way of reaching the Yukon." Ultimately, three routes emerged as most popular, if not well known—the Chilkoot Trail, the White Pass and the Edmonton Route. There was also an All Water Route, up the Yukon River via the old Russian port of St. Michael's in Alaska. Boats plied

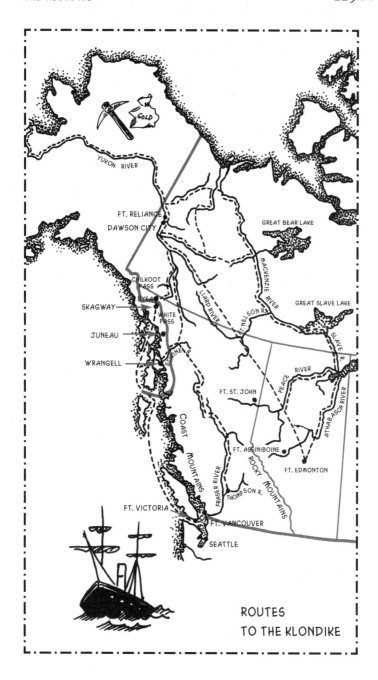

that thoroughfare before the rush, so it too was well-known. It was also the longest route and the most expensive, so only the few well-to-do bound for the Klondike actually used it.

Most Klondikers found themselves ferried up the Lynn Canal, where they came smack against the most challenging stretch of their journey, the Coast Mountains. The miners disembarked at either Dyea or Skagway, where they made tracks for Lake Bennett, British Columbia. Arrival at Lake Bennett meant that they still had about 500 miles to cover before they could dip a pan. Except for Miles Canyon and the rapids that followed it, those 500 miles were a Sunday stroll compared to the 30-odd miles between the chosen port on the Alaskan panhandle and the lake. It was some of the most difficult territory imaginable, and even a detailed description could not possibly have prepared the miner for it.

From Dyea the miner followed the 28-mile long Chilkoot Trail. For the first half of the hike, the miners struggled skyward for 3700 feet, the last stretch at a steep 35 degrees, until they crested the Chilkoot Pass. There they entered Canada and were greeted by the North-West Mounted Police. The Mounties' presence was for the protection of the miners. They knew that the Klondike was an isolated region, where supplies were limited and expensive. They didn't want anyone going inland who wasn't prepared. Preparation was defined as 1000 pounds of supplies, everything from flour and beans to candles and soap, not including clothing and equipment. The requirements meant that a miner had to make an average of 40 trips up the Chilkoot Pass. The undertaking was eased in the winter when 1200 steps, the Golden Stairs, were carved into the snow. In spring and fall, however, the trail was mud and one step forward might literally mean two steps back and a nasty fall against

Competition for prospectors and their business was fierce in the early days of the Klondike Gold Rush. Two towns that had sprung up and flourished in the wake of the discovery of gold were Dyea and Skagway. Just 10 miles apart, each was situated close to a route over the Coast Mountains. Dyea was the trailhead for the Chilkoot Pass. Long before it became a prospectors' staging point, Dyea was a seasonal fishing camp. But by the spring of 1897, Dyea was booming. While the Dyea Trail was one of the most popular routes into the Klondike, it was not without its challenges. The trail was littered with boulders deposited by retreating glaciers making passage demanding and time-consuming. Prospectors with money were able to ease their efforts by hiring Natives as packers. The Natives weren't always willing participants in the gold rush. The Chilkoot people resisted the initial wave of prospectors, and only relented when a U.S. gunboat steamed up the Lynn Canal with Gatling guns blazing.

one of the many boulders that littered the trail. A miner could expect to spend three months lugging his gear to the summit.

The Chilkoot Pass took on a life of its own because of the time required to traverse it. The Scales, where goods were packed and repacked for the actual climb, was at the base of the pass. Entrepreneurs built an aerial hoist there in 1898. Some three miles from there was Sheep Camp. Its population ran as high as 8000 in the winter of 1897 to 1898 and consisted of those on route, those who were trying to decide whether it was actually worth continuing on and those who were making a fortune from both. There was a hotel and a restaurant and a deckful of professional gamblers. Other criminal activities went far beyond the use of a marked deck, weighted dice or shell games. It was a place where might proved right as was evident in the daily robberies and murders. Among the honest folks looking to make a living, the ones who made the real money were those who transported cargo. Given that nothing but a man could climb the pass, everything including gear, sledges, and dogs (those with horses had long since abandoned them) had to be carried. They charged by the pound and the distance and most made a good $25 a day, though it's doubtful any fee could compensate for the challenge of hauling up a piano strapped to one's back. For most of the transporters, it was a short-lived career followed by a lifetime's misery of weakened ankles.

Some 20,000 miners negotiated the Chilkoot Trail. It's not known how many thousands more were bested by it. On learning of the demands of the pass, many miners turned to the nearby trail out of Skagway. It took the miner through the White Pass and, though the route was a few miles longer, it was also a 1000 feet lower and not nearly as steep. Ten thousand miners eventually hiked it, but the Skagway Trail came with its own tests. The first was encountered in the

Shown in the foreground is Soapy Smith, photographed in one of his saloons in Alaska. There were many who earned their fortune bilking miners, but Soapy Smith's operation was one of the most lucrative. Its tentacles reached into all facets of life in the Klondike, providing those "extras" for miners who wanted to relax after long days and nights at their claims. Most of the troubles in Skagway were said to be caused by the activities of Soapy Smith's gang. Major Sam Steele of the NWMP was determined that they not be allowed across the border into Canada. To that end, he installed Maxim machine guns at both border crossings and gave orders to turn back Smith's men. Within weeks, Soapy's gang was out of the picture, though it continued to thrive in the towns along the Alaskan panhandle.

town itself. It was a true gold rush town with 5000 residents by early 1898. The Mountie Sam Steele described it as "the roughest place in the world." With precious little in the way of a police presence, the man who controlled most of the crime in the town, Soapy Smith, discovered his own gold strike. The tentacles of his operation that was 150 men strong reached into every dance hall, saloon and brothel, all of which did dizzying business. Whether it was marked cards, watered-down booze, or a bruising encounter with a blackjack, few in Skagway didn't contribute to filling Soapy's pockets. He wasn't all brass knuckles though, as demonstrated by his telegraph office con. Soapy charged $5 a message, a deal in the minds of many prospectors desperate to send word to worried family members. Those who used the service invariably received a wire from home urgently requesting money for some emergency. Soapy wired that as well, right into his coffers. Had prospectors taken the time to look, they would have discovered that no telegraph wires left the town.

There was good reason to get out of Skagway with some speed. The trail that awaited was not necessarily a better place to be, though it probably started out as such. Because the climb to the White Pass was not as steep, it was more suited to pack animals, so many prospectors paid top dollar for horses and most discovered that their eyes for horseflesh were cataract-covered. Skagway was full of horses that should have been put out to pasture and even those strong enough to make the journey were so ill-used by their owners that they were soon ready to be put out to pasture. It's difficult to blame the prospectors. Many were city folk who knew nothing about animals and they were confronted with transporting half a ton of supplies, so it was natural enough to pile the packs high on the animals' backs.

The horses, and occasionally mules and goats, suffered terribly. The scene is best described by the pen of the

Tent towns sprang up all over the Skagway River valley during the gold rush. The dash to stake claims in the Klondike led to overcrowding on the White and Chilkoot passes, leaving many prospectors stranded and restless. Many remained in Skagway, waiting for their next window of opportunity. Others chose to stay when they realized that they had neither the endurance nor the means to make the perilous trek. Some, who felt that they would never get there in time to stake their claims, never moved on. Squatters set up tents all across William Moore's homestead. His claims to the property were ignored, and self-appointed government officials usurped Moore's legal rights, establishing the tent city of Skagway. It was a foul-smelling place whose muddy streets were lined with horse manure and sewage and where the law was non-existent. Moore was eventually awarded 25% of the value of the lots within his original town site.

adventure writer Jack London, who was an eye-witness to the misery. "The horses died like mosquitoes in the first frost and from Skagway to Bennett they rotted in heaps. They died at the rocks, they were poisoned at the summit, and they starved at the lakes; they fell off the trail, what there was of it, and they went through it; in the river they drowned under their loads or were smashed to pieces against the boulders; they snapped their legs in crevices and broke their backs falling backwards with their packs; in the sloughs they sank from fright or smothered in the slime; and they were disemboweled in the bogs where the corduroy logs turned end up in the mud; men shot them, worked them to death and when they were gone, went back to the beach and bought more. Some did not bother to shoot them, stripping their saddles off and the shoes and leaving them where they fell. Their hearts turned to stone—those which did not break—and they became beasts, the men on the Dead Horse Trail."

The Mounties often took care of those animals suffering too much to continue on by mercifully shooting them when they reached the Canadian border. Some animals apparently didn't wait to be shot. Tappan Adney, a local correspondent for the *Harper's Illustrated Weekly*, wrote of a horse that walked off the edge of Porcupine Hill. All who saw it swore that the animal had committed suicide. Soon suffering and hardened hearts would be a thing of the past. In 1898, construction began on the White Pass and Yukon Railway. By February of the following year, track had been laid to the pass and five months later it reached Bennett Lake. With its completion, the best route to the Klondike was obvious. Dyea died as quickly as it was born but, by then, the rush had already begun to slow.

For those daunted by the prospect of hauling up backpack after backpack of goods up the Chilkoot Pass, the White Pass was much more attractive. Although longer, the White Pass allowed for the use of pack animals. With the North-West Mounted Police sitting atop both the White and Chilkoot passes ensuring that the men passing through had adequate supplies, pack animals were critical to the speed and success of any prospector's quest. Of course, adequate to the Canadian government meant close to 1000 pounds of goods. A prospector working without animals or help walked 80 miles for every single mile of goods moved. Little wonder, then, that in the rush to reach the gold fields, prospectors placed far more weight on their backs than the animals could handle, resulting in the deaths of countless beasts on the sometimes inhospitable terrain.

A would-be prospector stood at the Liard River watershed and, as his eyes scanned what lay before him, his shoulders fell and he slumped to the ground. He had been months (was it seven or nine?) on the trail from Edmonton to Dawson. He had fought the brush that had conquered ill-defined and seldom used Native trails. He had struggled to follow other poorly marked scratches and he considered that only luck had seen him through. He had braved sucking muskeg and endured bloodthirsty mosquitoes. He had slept under trees and stars and through rain and sleet. Months ago his first pack horse had died. One at a time the rest had followed, a slow punishment. When the prospector was forced to carry what he could of his two years worth of supplies, he had a new respect for the animals and a terribly guilty conscience. He didn't carry the gear far. He didn't have the strength. His gums bled and livid spots patched his body, but he wasn't thinking straight enough to know that he had scurvy. Over the past month he had started to abandon supplies, so the trail was marked like the bread-crumbed route of a fairy tale. But this trek was no child's story; it was a nightmare. As he looked at the foaming Liard and the canyons that lay ahead, he threw off his remaining packs from his back. He pulled something out, scribbled on it, and tacked it to a tree. Then he walked into the river and allowed it to sweep him away. A passerby would later read his epitaph, "Hell can't be worse than this trail. I'll chance it."

There were two routes that one could take from Edmonton to Dawson. The easier of the two followed the northern waterways along the Mackenzie River for some 2600 miles. Though it took as long as a year and a half to complete, 874 men and 11 women set out to travel it.

About two-thirds of them actually reached their destination. The overland Edmonton Route, presumptuously described as a "route," to the Klondike went through Fort Assiniboine, Peace River Crossing and Fort St. John. Many miners took a quick detour into the old Cassiar gold field before they reached the Liard River, just south of the Yukon. At that point, the prospector was about halfway to his destination. Once the canyons, white water and portages of the Liard were negotiated and the miner arrived at Pelly Banks, the route followed the Pelly River until that waterway joined with the Yukon River that led to Dawson. In total, the route was more than 1400 miles. When the rush was at full flow 766 men and 9 women set out for the Klondike along the Edmonton Route. The amazing statistic is that 160 of the men (and none of the women) actually made it.

The nagging question associated with the overland Edmonton Route is why a man or a woman would choose to take it. *The Klondike News*, which spoke with some authority on the routes to the northern gold country, advised against it. "The Edmonton route is out of the question at present [April 1898] for anyone taking in an outfit [and all did], as it involves long portages between rivers and lakes and hundreds of miles of travel through an unknown country. It would take fully six months to reach Dawson this way." If the reporter who wrote the story had doubled that estimate, he would have been more accurate. There was nothing to recommend the trail from Edmonton except for one very important fact, it was an all-Canadian route.

A quick glance at the Klondike suggests why an all-Canadian route was loudly and widely touted. While the gold fields were actually in Canadian territory, the most popular access routes took prospectors through the United States. Their ignorance of the region was such that many prospectors didn't even realize the diggings were in Canadian territory. This caused certain administrative problems, mostly dealing with unanticipated Canadian regulations (licenses, royalties, customs, supplies). Patriotism was also an occasional burr under prospectors' saddles. Many Canadians, and to a lesser extent Britons, wanted to get to the Klondike without having to travel through American territory. Government officials also saw the route as a way to more solidly link the north to the southern part of the nation. Certainly the most significant reason for the promotion of an all-Canadian route was economic. Canadian businesses bristled as they watched *their* gold field being exploited by American interests in Skagway and Dyea, but also in the larger southern cities like San Francisco and Seattle. Simply put, they were envious. Shouldn't Canadian businesses be the ones to profit? The answer seemed obvious. J. W. Dafoe, the prominent newspaper editor of the *Winnipeg Free Press*, summed up the prevailing sentiment when he wrote that to question the Edmonton route "was regarded as a species of treason."

There were a few all-Canadian routes promoted, including the Ashcroft (Telegraph Trail) that began on the Thompson River and hugged the eastern edge of the Rocky Mountains through British Columbia. Fifteen hundred prospectors traveled it. The shorter Stikine Trail actually cut through the Alaska panhandle but, because it followed the Stikine River, there was no need to stop in American territory. There may well have been thousands who set off towards this route, but they were all disappointed because the railroad they expected to find was never built. Instead they ended up in Wrangell, a violent place controlled by Soapy Smith's gang. Most returned home or departed in search of another route. As devastating as the Edmonton Route was, it was the most successful, if not the most well-known of the all-Canadian routes.

Pictured here are miners waiting to register their claims with the claims office in Dawson, Yukon. Joseph Ladue, a local trader and entrepreneur, guaranteed his own gold strike when he staked out a town site at the junction of the Klondike and Yukon rivers. He called it Dawson, after the Canadian geologist George Dawson. He then built a warehouse and a sawmill for the construction boom he sensed was inevitable. And the boom did come. It was slow going at first as the news of $200 pans ambled to Circle City. By the summer of 1897, this first wave of the rush had crested at about 4000. It may be said that he was instrumental in starting the Klondike rush since he financed Robert Henderson who then passed on information about his find to George Carmack.

Although there is some debate about whether Edmonton businesses began advertising their town as a natural departure point for the Klondike before the arrival of the first Klondikers, there is no doubt that they were quick to jump on the bandwagon. A local newspaper, the *Edmonton Bulletin*, led the leaping. In August 1897, it produced a special Klondike issue headlined "To The Golden Yukon The All Canadian Route" that sang the praises of the Edmonton Route, while criticizing the long distances and difficulties of the other routes. Of course, it was noted that all the necessary goods could easily be purchased in Edmonton. The local retailer McDougall & Secord soon produced its own *Guide to the Gold Fields* that detailed those goods and their costs for a one-year supply. The list consisted of three categories, groceries (30 items for $91, but that had to be doubled because two-year supply was required), hardware (37 items for $62) and clothing (24 items for $91). Horses and tack were also necessary. It was advised that a man buy 10 horses ($25 each and another $50 to saddle them), eight to pack, one to ride and a spare. A Klondiker was looking at over $600 before he could get out of town.

Given those numbers, it's not surprising that Edmonton enjoyed its own gold rush boom. Before the discovery of Klondike gold, the town's population was about 1200. As it became known as the "Gateway to the North" that number swelled with those who arrived bound for gold country, though few stayed for long. Nevertheless, enough remained so that the town took on the appearance of a poorly organized army camp. Tents rolled back from the shores of the North Saskatchewan River. The main street, only five blocks long, was congested with bulging sledges ready to be moved out. Ladies had to step delicately to avoid the piled evidence of previously unseen numbers of trail animals. Hotels, saloons and restaurants naturally did well, and local businessmen not directly associated with the Klondikers also turned their energies to cashing in on the rush. A blacksmith invented an improved travois that reputedly could replace pack saddles and a brewer developed what he advertised as "evaporated potatoes."

It would be misleading to paint too greedy a picture of Edmontonians. For the most part, they were only doing what the residents of other cities did—boosting their town. They did it through the common enough practice of exaggeration—a good, well-marked trail, 90 days en route at the outside. Soon folks were cursing the "Gateway to the North" and, as the sordid details of the route became widespread, competing interests further blackened Edmonton's name. A series of headlines in the *Victoria Colonist* (a promoter of the Stikine Route) demonstrated how dark it could get. "Died For Gold. Lured From Edmonton Across Trackless Wilderness and Abandonned [sic] to Solitary Death in Most Appaling [sic] Form. Survivors of the Spectral Trail Reach Civilization with Harrowing Tales of Named and Unknown Dead. Men Cruelly Abandonned [sic] to Certain Doom Write With Touching Simplicity of Visions of Home Seen by Last Flicker of Vital Spark."

And it would be just as misleading to suggest all that the Klondikers were crazed fanatics, set on getting to the gold fields despite what were increasingly known as the perils of the journey from Edmonton. Certainly some, like Viscount Avonmore, were well described as such. He was the noblest member of the Helpman party that arrived from Ireland around Christmas of 1898. His companions included three colonels, three captains and two physicians. Perhaps an indication of their preparedness, or lack of it, were the 75 cases of vintage champagne they brought to town. Avonmore was soon known as Lord "Have One More." Of the group, one died, three were seriously injured and none reached the

Klondike. But there were those for whom Edmonton provided an opportunity for sober second thought and, with that, a return home. One such example is a lady who left a poem in a hotel room. It's rather appropriate that the poetess was anonymous.

> *To rush off to the Klondike there is surely no need,*
> *Where thousands may fail while a dozen succeed,*
> *When nuggets, as precious as specimen shown,*
> *At home may be found on a claim of your own.*

༄

Those Klondikers who traveled through Skagway or Dyea and learned that there was good cause to fear for their lives and property there, were surely shocked to learn that they need be concerned about neither once they crossed the boundary into the Yukon, where the lawlessness of American territory was transformed into an orderliness that bordered on the zealous. The attitude that Yukon justice should ensure that a miner could concentrate on working a claim was little different than the prevailing sentiment during the Fraser River and Cariboo rushes decades before. There was one major difference between the Yukon and British Columbia—the North-West Mounted Police. The Mounties were given the tough assignment of ensuring that flotsam and jetsam caught in the current of the gold rush rolled no farther than the border. Superintendent Sam Steele mostly oversaw the scarlet-jacketed efforts that met the challenge. His will proved far less malleable than the ore men came seeking.

A forty-niner who arrived in the Yukon in the years leading up to the rush of '98 would have felt pretty comfortable and with good reason. The Yukon was a lot like the old California before the North-West Mounted Police arrived in 1894. Miners kept their own order through meetings where they established laws and decided, with a show of hands, the fate of those charged with an offence. As grass-roots as the process was, it was a suspicious operation, not just because it was thoroughly un-British. As often as not, the penalty assessed was a fine that quickly paid for a round or two for those who sat in judgement. It seems no coincidence that the local saloons served as courthouses. Eventually a miners' meeting fined a man who wasn't willing to sit still for their brand of justice and who had the connections to do something about it. John J. Healy was well known to the Mounties. For a time he owned Fort Whoop-Up, the notorious watering hole in the southwestern corner of the Canadian prairies. He had also been a lawman in Montana. When he wrote to Sam Steele and informed him that locals were demoralizing Natives with the sale of whisky and thereby setting the stage for violence, Steele listened and hurriedly dispatched Inspector Charles Constantine. In 1895 he was joined by a further detachment of 20 men. Constantine's first point of business was to abolish the miners' meetings.

In the fall of 1896, Constantine advised his superiors that a new post with upwards of 25 men should be built in Dawson. He noted that there were already 350 claims made there, which represented 2000 miners plus hundreds more camp followers. His post was more than 50 miles away at Fortymile, too distant to effectively monitor the new diggings. Reinforcements, including Sam Steele, arrived in the early winter of 1897. Constantine was assigned to Dawson and those who accompanied Steele laid down the law at Lake Bennett. Since it was at the end of the Skagway Trail and just past the end of the Chilkoot Trail, Bennett was, for most stampeders, the Canadian gateway to the Klondike.

It was 10 PM, although the daylight gave no hint of it, and Sam Steele still had a couple of hours to put in. He had been at work since 5 AM, a routine he had settled into when he first arrived at Bennett. Except for meals and a stretch of his legs, paperwork, interviews and settling disputes had kept him in his office all day, yet another routine imposed by the demands of the Klondike. A bed in the corner kept him there all night. Steele saw little of Bennett.

"Corporal, have you the latest statistics on incoming prospectors?"

"Here they are, sir," replied his clerk, Corporal Tennant.

As Tennant passed the document to Steele, a shot rang out. Since the arrival of Steele it was a rare sound in Bennett.

"See what it is and report back to me," ordered Steele with a nod of his head towards the door.

"Sir."

A few minutes later Tennant returned. With him was a clean shaven, well-dressed man in handcuffs.

"He's no miner," Steele said to Tennant.

"Folks say he's with Soapy Smith's gang."

"Indeed. A con man then."

"I got my rights!" shouted the man. "I'm an American citizen! You can't lock me up here."

"Sir, the last thing I want to do is offend our good American neighbors. Seeing that you're one, I'm going to be lenient," replied Steele.

The man gave a smug nod. Steele looked to Tennant.

"Corporal, confiscate his goods." He turned to the con man. "You, sir, have half an hour to get out of town."

Lenient! He was still sputtering as Tennant marched him out of the room.

Sam Steele was direct, efficient and about as complicated as a slipknot, but the most revealing aspect of this story is

There were two reasons why the gold rushes in Canada were more orderly than those of the U.S. The first reason was the efforts of James Douglas and his appointed judge, Matthew Baillie Begbie, who let miners know right from the start that they were on British land and that they would obey the laws of Britain. The second reason was the North-West Mounted Police, who patrolled the trails and towns of the Yukon keeping order. But the Mounties did much more than police. They ensured that prospectors had enough supplies to sustain them through their time in the north, thereby preventing much hardship. They made certain that the sick were cared for and that those less fortunate were buried. And, as Superintendent Sam Steele reflected, they did it all "without one well-founded complaint against us."

that he wasn't trying to be funny. He was almost 50 when he arrived in the Yukon and he brought with him a wealth of experience. For 30 years he had been employed as a lawman, some of it with the military, but most with the North-West Mounted Police. He was part of the force's historic march west in the mid-1870s and a decade later he was a superintendent. He was in command of the Macleod district in southern Alberta when he was assigned to the Yukon.

It was −30°F on February 14 when Steele arrived in Skagway where the Mounties had an administrative office that provided incoming miners with necessary advice. Posts were being built at both the Chilkoot and White Passes to serve as custom offices and to guard the passes and, despite the fact that a northern storm raged, Steele set out the very next day for Dyea to begin an inspection. He spent the night in Dyea and was on the Chilkoot Trail in the gloomy darkness of the next morning. The route was desolate and the cutting wind was such that Steele and his companions were forced to take regular shelter behind trees and sleighs. They passed through Sheep Camp the next morning and continued onto the Scales, which they discovered the storm had covered in a blanket of snow. Eventually they found a tunnel and, following it into a snowdrift, they eventually came to two previously hidden tents. The covering snow provided such insulation that it was the warmest Steele had been since arriving in Skagway. He did not have to climb the pass the next day, which was just as well because the storm continued; the Mountie assigned to the post on the Chilkoot slipped down into the Scales to inform Steele that the post was ready. With little more than a curt nod of his head, Steele returned to Skagway.

While Steele's office was being readied in Bennett, he remained at the Skagway office. He quickly learned that it was not a place where an honest man might live with any confidence. At night, cries of murder and pleas for help mingled with the songs and clamor of the saloons and dance halls. On more than one occasion, Steele was awakened by gunshots, the lead from which occasionally found themselves embedded in the walls of his quarters. The place was simply chaotic, and Steele figured that the only people safe in town were the Mounties themselves. Most of the troubles were caused by the activities of Soapy Smith's gang and Steele was determined that they would not be allowed across the border. To that end, he installed Maxim machine guns at both border crossings and gave orders to turn back Smith's men and, with that, end the gambling and robbery that was so common it affected just about every Klondiker. Literally within weeks, Steele could confidently state that "there was no danger of Soapy Smith or his gang; they dared not show their faces in the Yukon…. [E]veryone went about his business with as strong as sense of security as if he were in the most law-abiding part of the globe."

Once at Bennett, Steele did much more than keep it safe, which actually proved to be the easiest of his responsibilities. The real challenge lay in protecting the Klondikers from the challenges that awaited them. Steele enforced the regulation that required each miner to have a year's worth of supplies; he saw it as a sensible and valuable requirement. He also put into effect other regulations designed to improve the likelihood that miners would arrive safely at Dawson. One of the greatest challenges faced by Klondikers was Miles Canyon and the White Horse and Squaw Rapids that churned just north of it. The foaming white waters regularly swamped or destroyed boats that attempted to pass, often with loss of life. Steele ordered that boats were to be evaluated by the Mounties to determine whether they were sound enough for the journey. Pilots were selected to steer the vessels and under no circumstances were women or children allowed to

One of the greatest challenges faced by Klondikers was Miles Canyon (shown here) and the White Horse and Squaw rapids that churned just north of it on the upper Yukon River. The five miles of foaming white waters regularly swamped or destroyed boats that attempted to pass, often with loss of life. Finally, Superintendent Sam Steele ordered that the boats be evaluated by the Mounties to determine whether they were sound enough for the journey. Pilots were selected to steer the vessels and under no circumstances were women or children allowed to be taken in the boats; they had to walk. Neither vessels nor lives were lost after the regulations were put into place.

be taken in the boats. They had to walk the five miles. Neither vessels nor lives were lost after the regulations were put into place.

Steele was pleased with the performance of the Mounties in Bennett. He later wrote, "More than 30,000 persons, everyone of whom had received assistance or advice, had passed down the Yukon. Over $150,000 in duty and fees had been collected, more than thirty million pounds of solid food...had been inspected and checked by us. We had seen that the sick were cared for, had buried the dead, administered their estates to the satisfaction of their kin, had brought on our own supplies and means of transport, had built our own quarters and administered the laws of Canada without one well-founded complaint against us. Only three homicides had taken place, none of them preventable, a record which should and...did give satisfaction to the government of the Dominion." Indeed, authorities were well satisfied. The corridors of power in Ottawa were witness to open acknowledgment of Steele's efforts, particularly as they related to the prevention of mass starvation and also to whispers that his diligence forestalled rumored American plans to slip into the region and take control under the guise of rescuing its citizens. Steele was subsequently appointed to the Council of the Yukon Territories, put in command of the North-West Mounted Police in the Yukon and British Columbia and dispatched to Dawson to replace Inspector Constantine.

How effective had Charles Constantine been at keeping the peace in Dawson? When he was reassigned in the summer of 1898, residents presented him with a silver plate embedded with $2000 worth of gold nuggets. No one doubted that it was the only gold he took from the Yukon. Like every mining town, Dawson had its saloons, dance halls, theaters and brothels, and with them gamblers, drunks, prostitutes and pimps. What differed in Dawson from other mining towns was that it had little of the crime associated with those activities. The peacefulness that characterized the community was due to the tireless and fearless work of Constantine, Steele and other Mounties whose names have largely been forgotten by history.

At its height in 1898, Dawson boasted a population of 30,000 and the majority were Americans. In that year there were 650 arrests. Less than a quarter were for crimes more serious than a misdemeanor and, of those, more than half were for prostitution. It's not difficult to figure out why it was such a serene place. When a fellow arrived in Dawson, almost his first act was to relinquish his revolver. Those who were unwilling to hand it over had it taken, or were turned around and sent back to where they came from. Needless to say, that one regulation almost eliminated bloodshed and the violent crime associated with it. To further reduce the chance of violence, Constantine and Steele had detectives investigate suspicious characters and infiltrate suspected gangs. If, per chance, an argument did get out of hand and it appeared that assault was inevitable, one only needed to take the advice of a resident, "If you get into trouble call a policeman.... The old American stall of self-defense just doesn't go."

When it came to vice, the Mounties (with the support of local governing officials) adopted a policy of regulation. Saloons and dance halls were licensed (and closed on Sundays, because there was actually no work of any kind allowed on the Lord's Day), gambling was permitted as long as there were no complaints about the fairness of the game, and prostitutes plied their trade as long as they underwent medical exams to verify their health. The Mounties cracked down on disorderly conduct and obscenity. The money raised from licenses and fines went towards the general upkeep of the city—sanitation and hospitals particularly

The Klondike

Although Dawson City benefited greatly from the discovery of gold and became the "Paris of the North," its success came at a cost. The Hän people, who had lived in and around the Yukon River for generations, found their way of life undermined and jeopardized by the rapid settlement of Dawson City. Although they too benefited from the new economy, they lost many of the resources on which they depended. The prospectors drove game from the valley, cut trees needed for fire and became competition at fishing sites. To accommodate both cultures, the Hän decided to move from their village, which sat on land eyed by rapidly expanding Dawson City, and relocate three miles down the river at Moosehide. Here, the Hän sold moccasins, gloves and moose meat to miners and were able to remain self-sufficient. Over time, however, resources became depleted and they suffered a fate similar to that of other Natives across the continent.

benefited. When news of this state of affairs made its way to Ottawa, there was public outcry. It appeared as if Yukon authorities were condoning immorality (and permitting activities that were illegal in the south). Of course, it was more a matter of effective management, but by the turn of the century, federal government officials had their way and clamped down on the offensive activities themselves.

Tappan Adney, the correspondent for the *Harper's Illustrated Weekly,* lived in Dawson in 1898. He was rarely a fan of the efforts of the local administration and regularly, and sometimes appropriately, criticized officials for corruption. But he had nothing but flowery praise for the Mounties. "The police control of the country was as nearly perfect as one could expect…. No city on the continent presented a more orderly appearance." In demanding that order, Sam Steele effectively wrote his own ticket out of Dawson. There was a minority in town who didn't like his authoritative ways and they were finally successful in convincing his superiors to have him transferred out of the Klondike. When he left, however, he could confidently state that he "stood up for the credit of Canada and the honor of the force to which I belonged, and it [was] no idle boast that at no time in its history did the police show to better advantage than during the trying years of 1898–99, when I commanded the fine officers and men on the Yukon." There was, as he declared with understated satisfaction, no place for a Soapy Smith in the Yukon.

As part of his assignment to cover the Klondike Gold Rush, the correspondent Tappan Adney spent some time at the diggings along the Bonanza and Eldorado Creeks.

There were few women who participated on the rivers' shores in the gold rushes; however, saloon owners did their best to attract miners to their establishments by providing certain entertainment. During the gold rush, saloons, gaming parlors and dancehalls ran six days a week but, by order of the NWMP, were closed on Sundays. And there were prostitutes in Dawson City, but there were also dancehall girls who were paid to flirt outrageously with the miners. Some of these girls put themselves up for sale as wives to the miners. One was offered her weight in gold and at 112 pounds fetched a reverse dowry of $25,000. The dancehall girl pictured here is called Snake Hips Lulu. It's easy to guess where and why she received this colorful sobriquet.

He was there in late November 1897. What he saw he described as a "strange, weird sight.... one never to be forgotten," one, in fact, he considered that the most inventive of imaginations would have been challenged to concoct. "The sun, like a deep-red ball in a red glow, hung in the notch of Eldorado; the smoke settling down like a fog (for the evening fires were starting); men on the high dumps like spectres in the half-smoke, half-mist; faint outlines of scores of cabins; the creaking of windlasses—altogether a scene more suggestive of the infernal regions than any spot on earth. It was hard to believe that this was the spot towards which all the world was looking. Little more than a year ago this wilderness, now peopled by thousands of white men, resounded only to the wolf's howl and the raven's hollow *klonk*. Well might one gaze in wonder, whether an old California miner or one who had never before seen men dig gold, for the world had seen nothing like this." To investigate this singular happening, a stroll along the creeks seems appropriate.

A short hike would reveal a great deal. The standard claim was 500 feet long, though they were sometimes shorter due to the placement of adjacent claims or a miner's imperfect measurement. Occasionally, a miner's poor surveying skills meant another man's fortune. The best illustration that size didn't matter in the Klondike was claim Two Above. A fellow named Dusel had staked and registered it, but when William Ogilvie, the Dominion Land Surveyor, took his chains to it, it was discovered to be too long. Ogilvie, as a government official, couldn't make a claim but he suggested to one of his chain men that he register the claim. Dick Lowe wasn't overly excited about the prospect. The excess proved to be just less than 90 feet at its widest, which wasn't much. But Lowe was a late arrival to the Klondike and there wasn't much left to stake. Nonetheless, he first prospected for another claim. Unsuccessful, he registered the piece next to Two Above. He then tried to unload it and was willing, in fact, to take $900 for it. There were no buyers, so he mined it himself. After eight hours work he had $46,000. Ultimately he mined a good half a million dollars, and Dick Lowe's Fraction became known as the richest piece of ground ever found.

From the fork of the two creeks, there were about 100 claims within a ten-mile stretch in each of the three directions. Many fires were visible along the creeks, hardly surprising given the season, but the purpose of the flames was not just to keep the miners warm. Fires also served to facilitate the excavating. Placer mining dominated and the miners concentrated on the gravel bars in the creeks, which presented problems. The creeks were free of ice for only a few months and the soil itself was in a state of permafrost. But the miners were nothing if not determined, and they met the challenge by using fires. As the ground heated and thawed, they scooped up the gravel, and the fire went deeper into the ground as buckets of gravel were removed. By 1899, this technique was refined. Steam was pumped out from engines into pipes and hoses. The head of the pipe or hose was reinforced so that it could be pounded into the ground. A hose that was left running for 6 to 12 hours allowed for the removal of 1 to 3 cubic yards of gravel. Because the ground was so hard, it was easy to tunnel without worrying about cave-ins and the use of steam rather than fire meant that asphyxiation was no longer a concern. In the spring, the recovered gravel was washed out and, if lucky, the gold was recovered.

The work meant long hours in uncomfortable conditions. In the winter, miners braved bone-chilling cold as they perched above their shafts and operated the windlasses. Before steam technology was introduced, they had to perform the operations teary-eyed, as smoke billowed up the shaft. As supplies began to dwindle, the miners' health suffered accordingly and there were more than a few cases of scurvy along the creeks. In summer they had to take advantage of the running water to operate their sluices, so sleep became something of a luxury. Blackflies and mosquitoes were a constant torment, severe enough to be cursed if a miner wanted to take the chance of having his mouth invaded. It was a hard life without any guarantee of a pay-off. Indeed, for most of those who arrived after 1897, after the Bonanza and Eldorado creeks and their tributaries were staked by sourdoughs (those who had seen a winter in the Klondike), there was nothing *but* hard work.

Because claims were few and far between after 1897, the opportunities for *cheechackos* (a Tagish word meaning newcomer) were limited. Some continued to prospect, moving into the hills that surrounded the rich creeks. Occasionally they even struck it rich, one such site being the White Channel on Cheechako Hill. The prospector who discovered it was Oliver Millet. Poor health forced him to sell his claim for $60,000 and it eventually brought the new owners half a million dollars. Those who had money, or good credit, could buy a claim or an interest in one. The most successful at that practice was Alex McDonald, the King of the Klondike, and one of its richest men.

McDonald came to the Klondike from Antigonish, Nova Scotia, by way of a long stop at the silver mines of Colorado and a brief go for the gold in Juneau, Alaska. His mining experience led him to believe that the best way to make money was to have others work his claims. He first acquired a 50% interest in Thirty Eldorado. Then he leased (the lay system as it was called) a section of it out, taking 50% of what the miners found. Within a month and a half he had made $16,000, not bad for an initial outlay of some

That women actually came to the Klondike when they so rarely joined earlier rushes was largely a result of society's shifting view of women. At a time when sexual equality was being discussed with great frankness and greater optimism, some women saw participation in the rush as a way to transform ideas into action. If men could do it, so could they. Most of the women ended up in Dawson. Only 1% of the miners were female and none were among those who struck it rich. More common were women who owned a claim; 3% of all Klondike claims were registered to women. For most of these women, it's likely that ownership was part of a family economic strategy. A miner could only hold one claim per creek, but a married miner could hold two and thereby double the chances of hitting pay dirt.

groceries. He parlayed that money into ownership of other claims. But there was nothing certain about what McDonald did; it was all speculation and he had as good a chance to go bankrupt as he did to become rich. His practice was to give only a down payment on a claim, with an agreement that he'd pay the rest in the spring. His dealings were based on the hope that when the winter's gravel was washed, he'd have enough to cover his debts. He did. At the peak of his success, he held interest in some 50 claims. No one knows for certain how much money he was worth, but it was well into the millions. The King of the Klondike, however, wasn't much interested in gold and his continued speculations ruined him. He died of a heart attack while chopping wood near his cabin.

Swiftwater Bill Gates also made a fortune using the lay system and a little deception. His nickname was as much a slight as anything else, bestowed because he claimed to have been a riverman on the Coeur d'Alene. If he was, it had been a poor paying job because most folks in the Klondike knew him as a beggar in Circle City. Desperate for a strike, he partnered up with a group of men to take a lay out on Thirteen Eldorado. The owner didn't want to work it and no one wanted to buy it because of the unlucky number. It looked as if the claim really was ill-favored until the group sank their seventh shaft and struck pay dirt. To keep it a secret, they told folks they were getting a lousy $.10 to the pan. They made an offer to the owner to buy Thirteen and he was happy to unload such a dud. They recovered the $45,000 they paid in six weeks and soon hired men to work the claim. As the money poured in, Gates made one more investment, a 30-foot square that yielded $85,000. His fortune secure, Gates took to spending it. Gambling, women and booze (not for drinking, but for bathing in) were his preferred vices. While he was a freewheeler, he was also something of a jealous man, who let his irritation show with a particularly black sense of humor. His girl, or so he thought, was a 19-year-old dancer, Gussie Lamore. She had one weakness, fresh eggs; but even the good wage of a dance hall woman was insufficient to buy them regularly because they were the rarest commodity in Dawson. On one occasion, Gates saw Lamore on the arm of another man. Gates turned on his heels and bought every egg in Dawson. He had them fried up, all 2200, and thrown to the dogs in the street.

There were others in the cabins along the creeks who were not as fortunate as either McDonald or Gates. Many nameless men turned to working the claims of others for a set wage. There was always a demand for employees because of a regulation that required a claim be worked or be forfeited. Miners who had more than one claim or who had done well on the one claim, generally hired others to do the manual labor. It was hard work, of which digging shafts and sifting gravel was only a part. Cutting wood was an ongoing activity—for the fires in winter, the sluices and rockers in summer and the cabins. The only good thing to be said for the work was that it paid well because demand usually exceeded the supply of workers. Of course, there was no guarantee that wages would be forthcoming. Most contracts stated that wages were "payable at wash-up" in the spring. If the mounds of gravel collected over the winter didn't yield much gold, laborers were often out of luck. A claim's owner could be taken to court, but it was a time-consuming process that rarely bore fruit.

A hike along the creeks also revealed the occasional woman. In 1898, there were about 1200 non-Native women (a huge jump from the 30 who were there three years before), 8% of the total residents. One's sex was no inoculation against gold fever and, through the discoveries of the 19th century, many women undoubtedly suffered from the same tremors that drove men into the unknown. That women actually came to the Klondike when they so rarely joined earlier rushes was largely a result of society's shifting view of women. At a time when sexual equality was being discussed with great frankness and greater optimism, some women saw participation in the rush as a way to transform ideas into action. If men could do it, so could they. Most of the women ended up in Dawson. Only 1% of the miners were female and none were among those who struck it rich. More common were women who owned a claim; 3% of all Klondike claims were registered to women. For most of these women, it's likely that ownership was part of a family economic strategy. A miner could only hold one claim per creek, but a married miner could hold two and thereby double the chances of hitting pay dirt.

Wives were key partners in other ways as well. Not only did they take care of the domestic concerns of a claim, they also took in work (laundry, sewing, cooking, berry picking) from neighboring claims. The importance of such work was magnified if the claim itself failed, which it often did. Those women with money also took to speculation, either by investing in claims or grubstaking a prospector. And, although it was an unusual sound, the attentive ear could hear babies crying and children playing. More than 200 families—man, woman and children—traveled to the Klondike. Three hundred more families joined their husbands and fathers after they had made a claim. There were also children born in the Klondike. However it may have happened, those women with children faced an even more strenuous life in the Klondike.

Many minority groups were drawn to Dawson, but few of them were found along the creeks. Most common, other than Americans and English Canadians, were French Canadians

The Klondike

As can be surmised from the photo, life was hard on the gold rush trail and as a result, most men left their families securely ensconced in their comfortable homes back east or down south. However, because of the shifting view of women in society, more and more women were joining their husbands at their claims in the Klondike. At the peak of the rush there were some 500 families prospecting in the Klondike. Mrs. Clarence Berry, a new bride, hiked with her husband to his claim only to find there a small shack with no doors, no windows and no facilities for bathing except the pan that also washed the gold from the gravel. When asked what advice she might give to other women about going to the Klondike she said, "Why...stay away, of course. It is no place for a woman."

and Scandinavians. The French Canadians, particularly, had a habit of gathering in enclaves when possible. They developed a strong social network, integrating *cheechackos* as they came, by sharing information about employment or lodging.

One group whose members were only occasionally seen along the creeks were the local Natives, the Hän. There were about 1000 Hän before the arrival of the Klondikers and they soon found themselves swamped, with disease reducing their numbers to a few hundred by the end of the rush. The Hän were semi-nomadic hunters and fishers, and their operations were centered on the Yukon River and the adjacent creeks. The Klondikers interrupted their traditional ways and depleted their food resources. Access to alcohol was further demoralizing. When Dawson was surveyed where the Hän had their habitual summer camp, they were relocated down river and required to give up any claim to their ancestral home. When the Hän asked for more than the 160 acres they were granted, the government declined, at least until they were certain there was no gold in the area. For the most part, the Hän did not interact closely with the miners. They were content to pursue their traditional livelihoods that often took the men on extended trapping and hunting expeditions. None owned claims, though they might occasionally work one on a casual basis. More common was their support of mining through the trade of meat and fish. Hän women played a more visible role as paid servants. It was uncommon for them to assume a more prominent role. Although there was intermarriage, it was not popular, as suggested by the reminiscences of the grandson of a Klondiker. "In no part of the world where isolated white men live among aborigines was the man who had a Native mistress held in greater disrespect than here."

A final point of interest along the Klondike creeks was Grand Forks. Most Klondikers headed into Dawson for

Belinda Mulroney (second from the right) is pictured here in front of the Magnet Roadhouse, which was 15 miles from Dawson and near the mines at Grand Forks. Mulroney owned a hotel called the Grand Forks and was one of the few women entrepreneurs of the day. Along with other hotel owners and service providers, she found her own personal gold mine in catering to the needs of the miners. During her time in the Klondike she owned shares in 10 lucrative mines because when she heard about a good claim, she invested in it. Said to be once the richest woman in the Klondike, she eventually lost all her fortune. But for a time she must have been well loved by the miners who certainly appreciated the homey touches the lady put into her operation.

supplies and whoop-ups, but where there were miners gathered, there was always opportunity to make money. Some folks simply couldn't wait to get to town to celebrate or commiserate. There were about 20 cabins in Grand Forks, and although miners used some, others served as hotels, saloons and restaurants. When it came to saloons, only the essentials were required—a couple of barrels and a few pieces of whipsawed lumber to serve as a counter with another on which to display the liquor behind the bar. One hotel, however, was a more refined establishment.

The Grand Forks was owned and run by Belinda Mulroney. A Pennsylvania coal miner's daughter, she arrived in Dawson with a cargo of cotton goods and hot water bottles that she sold at a profit of 600%. She then set up an operation in Dawson for a while, but she soon sold out, preferring to be closer to the mines where she figured the real money was. Amidst mocking and disbelieving residents of Dawson, she hauled her own lumber to Grand Forks. For $3.50 a meal, customers ate from china dishes set on clean table cloths. They could eat and stay overnight on the second floor for $12. Mulroney was a smooth operator. She listened to the scuttlebutt, and there was plenty of it in her place because its popularity drew miners from around the Klondike. When she heard of a good claim, she made an investment. By the summer of 1898, there were over 1000 residents in Grand Forks and she sold the hotel for $24,000. By 1899, she had an interest in ten claims. Although she eventually lost much of her gold rush fortune, there was a time when Mulroney was reputed to be the richest woman in the Klondike.

Pictured here on the left is the Dewey Hotel. During the Klondike Gold Rush, many hotels sprang up at Grand Forks in the Yukon Territory between Eldorado and Bonanza creeks to provide miners with the amenities of home. The early hotels were simple chinked log structures thrown up wherever miners congregated. Later on, these places offered an opulence that could only be found with well-established hotels in New York and San Francisco. Miners who had gold to spend did so liberally and often went home with no more than they came with, whereas hotel owners benefited from the miners' willingness to spend lavishly.

When the curtain rose on the 20th century, it closed on the Klondike. But, while it was center stage, it held the attention of the world. Some 40,000 stampeders were swept north. Most never stood calf deep in the cold Bonanza, the Eldorado or their tributaries, but those who did took more than 33 million ounces of gold from creeks and river beds. It was more than enough to go around, and around it went. If only one thing could be said with certainty about miners, it's that very few of them were tight with their money. It was a characteristic that often caused destitution in later life, but while miners were spending their gold dust, the lights of Dawson glittered.

It wasn't to last. In 1898, Dawson had as many as 30,000 residents. A year later there were 20,000 and by 1901 there were less than 7,000. With the departing residents went much of the spirit that characterized the gold rush. The day when eccentric prospectors would stumble into town weighed down by heavy sacks of gold and throw a poke on a bar or at the feet of a dance hall girl were gone. They were replaced by men who worked the large electric-powered dredges that plied the rivers or the hydraulic operations that continued to dig shafts. There were still some prospectors in the Yukon, but they had mostly moved north in search of the Mother Lode that fed the creeks. It was in vain because the lode was never found.

The Klondike rush was a well-ordered affair. Mining licenses were required, lots were officially surveyed and the government collected a 10% royalty on all gold discovered. There were those who objected to such constraints, perhaps not surprisingly given that some four-fifths of the Klondikers were foreign. That objections were never a problem was mostly due to the steadying influence of the North-West Mounted Police. Their dedication might be evident in the fact that they were responsible for transporting some five tons of gold to Seattle during the rush. None of it was ever lost. For their efforts, the men were paid $1.25 a day for the job. Their presence ensured that miners could pan and dig and not worry about protecting their fortunes.

It wasn't just the staggering amounts of gold, the well-ordered nature of the rush, or even glittering Dawson that made the Klondike Gold Rush one of the biggest happenings of the 19th century. It was the characters who stampeded north and their experiences that gave it a unique flavor. The stories of George Carmack, Skookum Jim, the King of the Klondike, Swiftwater Bill and Belinda Mulroney are all well-known parts of the legend. But there were plenty of other nameless folks whose stories added spice to the mix. When Sam Steele was patrolling Lake Bennett, he was astounded to encounter a couple on their honeymoon. For all those who died on the Edmonton Trail, there was at least one baby who was conceived and born during the long journey. One miner sold his claim for $600, only to discover it was the widest pay streak in the country, running the full 500 feet. The new owners gave him 75 feet of it. And, in front of one cabin along the Eldorado Creek sat two containers, one a coal-oil can full of gold and the other a bottle of whisky. They were joined by a sign, "Help Yourself." A person who lives long enough will claim to have seen just about everything, but when the phrase "Hurrah for the Klondike" rolled off peoples' lips, just about as much could be seen within the space of only a few years.

Notes on Sources

A book of this nature draws on many sources, the most valuable of which are listed below. The events described in this book are true to the sources and fictionalized as little as possible.

California
Caughey, John W. *The California Gold Rush*. Berkeley: University of California Press, 1948.
Hill, Mary. *Gold: The California Story*. Berkeley: University of California Press, 1999.

Cariboo
Downs, Art. *Wagon Road North*. Surrey, BC: Foremost Publishing Company, 1960; revised, 1973.
Skelton, Robin. *They Call it the Cariboo*. Victoria: Sono Nis Press, 1980.
Wright, Richard T. *Barkerville: A Gold Rush Experience*. Williams Lake, BC: 1993; revised edition, 1998.

Klondike
Adney, Tappan. *The Klondike Stampede*. Harper & Brothers, 1900 (reprint ed. Vancouver: UBC Press, 1994).
Backhouse, Frances. *Women of the Klondike*. Vancouver: Whitecap Books, 1995.

Berton, Pierre. *Klondike: The Last Great Gold Rush, 1896–1899.* Toronto: McClelland & Stewart, 1958 (reprint ed. 1972).

Porsild, Charlene. *Gamblers and Dreamers.* Vancouver: UBC Press, 1998 (reprint ed. 1999).

Fraser River

Patenaude, Branwen. *Trails to Gold.* Victoria: Horsdal & Schubart, 1995.

Sterne, Netta. *Fraser Gold 1858! The Founding of British Columbia.* Pullman: Washington State University Press, 1998.

Black Hills

Parker, Watson. *Deadwood: The Golden Years.* Lincoln: University of Nebraska Press, 1981.

———. *Gold in the Black Hills.* Norman: University of Oklahoma Press, 1966.

General

Fetherling, Douglas. *The Gold Crusades: A Social History of Gold Rushes.* Toronto: University of Toronto Press, 1997.

Martinez, Lionel. *Gold Rushes of North America: An Illustrated History.* New Jersey: The Wellfleet Press, 1990.

Morrell, W.P. *The Gold Rushes.* London: Adam and Charles Black, 1940.

IF YOU ENJOYED *GOLD RUSHES*, YOU'LL LOVE THESE LEGENDS FROM FOLKLORE PUBLISHING...

Kootenai Brown
by Tony Hollihan

John George "Kootenai" Brown could boast countless adventures: serving in the British Army in India, hunting buffalo on the plains of Manitoba, riding Pony Express for the U.S. Army and fighting as a staunch conservationist in the region that eventually became Waterton National Park. Follow along in this exciting account of Kootenai Brown's incredible exploits.

ISBN 1-894864-00-X • 5.25" x 8.25" • 256 pages
$10.95 US • $14.95 CDN

Sitting Bull in Canada
by Tony Hollihan

This book recounts the story of Sioux chief Sitting Bull's retreat into Canada after the Battle of the Little Bighorn. The story centers on the friendship that developed between the fierce warrior and the celebrated Mountie, Major James Walsh.

ISBN 1-894864-02-6 • 5.25" x 8.25" • 288 pages
$10.95 US • $14.95 CDN

Great Chiefs, Volume I
by Tony Hollihan

Chronicled here are the lives of famous native chiefs and warriors who grappled with the increasing encroachment of European settlers in the West. The author dynamically brings to life these remarkable leaders and the means they adopted in a desperate bid to protect their people.

ISBN 1-894864-03-4 • 5.25" x 8.25" • 288 pages
$10.95 US • $14.95 CDN

Great Chiefs, Volume II
by Tony Hollihan

Tony Hollihan weaves more spell-binding tales of the courageous chiefs and warriors of North America's western tribes who battled valiantly against the growing tide of European settlement on their ancestral lands.

ISBN 1-894864-07-7 • 5.25" x 8.25" • 288 pages
$10.95 US • $14.95 CDN

Look for books in the Legends series at your local bookseller and newsstand or contact the distributor, Lone Pine Publishing, directly. In the U.S. call 1-800-518-3541. In Canada, call 1-800-661-9017.